# A Biometrical Study of Egg Production in the Domestic Fowl
## Variation in Annual Egg Production

### by US Dept. of Agriculture

### with an introduction by Jackson Chambers

# Self Reliance Books

Get more historic titles on animal and stock breeding, gardening and old fashioned skills by visiting us at:

http://selfreliancebooks.blogspot.com/

# *Introduction*

I am pleased to present yet another title on Poultry.

The work is in the Public Domain and is re-printed here in accordance with Federal Laws.

As with all reprinted books of this age that are intended to perfectly reproduce the original edition, considerable pains and effort had to be undertaken to correct fading and sometimes outright damage to existing proofs of this title. At times, this task is quite monumental, requiring an almost total "rebuilding" of some pages from digital proofs of multiple copies. Despite this, imperfections still sometimes exist in the final proof and may detract from the visual appearance of the text.

I hope you enjoy reading this book as much as I enjoyed making it available to readers again.

Jackson Chambers

1

2

# LETTER OF TRANSMITTAL.

U. S. DEPARTMENT OF AGRICULTURE,
BUREAU OF ANIMAL INDUSTRY,
*Washington, D. C., September 29, 1908.*

SIR: I have the honor to transmit herewith and to recommend for publication in the bulletin series of this Bureau a manuscript entitled "A Biometrical Study of Egg Production in the Domestic Fowl. I. Variation in Annual Egg Production."

This paper is the first in a proposed series dealing with a biometrical study by Drs. Raymond Pearl and Frank M. Surface of the results obtained at the Maine Agricultural Experiment Station in breeding for egg production during the past nine years, the work during the past three years having been done in cooperation with this Bureau. The subjects to be discussed in the several parts are briefly outlined in the introduction.

While the investigation is the joint work of Doctors Pearl and Surface, Doctor Pearl is responsible for the actual writing of the whole paper.

Very respectfully,

A. D. MELVIN,
*Chief of Bureau.*

Hon. JAMES WILSON,
*Secretary of Agriculture.*

# CONTENTS.

|                                                                         | Page. |
|-------------------------------------------------------------------------|-------|
| Introduction                                                            | 9     |
| Material                                                                | 11    |
| Statistical methods used                                                | 20    |
| Variation in egg production in Barred Plymouth Rocks                    | 24    |
|     Analytical discussion of variation in egg production | 30    |
|     Changes in egg production between 1899 and 1907  | 38    |
|         Changes in mean egg production | 39 |
|         Changes in the variability of egg production | 43 |
|         Changes in other constants | 46 |
| Variation in egg production in White Wyandottes                         | 49    |
| Egg production in other breeds of fowls                                 | 50    |
| The influence of certain housing conditions on annual egg production    | 58    |
| Egg production in the second laying year                                | 63    |
| Selection and egg production                                            | 67    |
| Summary of results and conclusions                                      | 73    |
| Appendix I                                                              | 76    |

# ILLUSTRATIONS.

|  |  | Page. |
|---|---|---|
| FIG. | 1. Variation in annual egg production, 1903–4 | 33 |
| | 2. Variation in annual egg production, 1904–5 | 33 |
| | 3. Variation in annual egg production, 1905–6, 50-bird pens | 34 |
| | 4. Variation in annual egg production, 1905–6, 100-bird pens | 34 |
| | 5. Variation in annual egg production, 1905–6, 150-bird pens | 35 |
| | 6. Variation in annual egg production, 1906–7, 50-bird pens | 35 |
| | 7. Variation in annual egg production, 1906–7, 100-bird pens | 36 |
| | 8. Variation in annual egg production, 1906–7, 150-bird pens | 36 |
| | 9. Change in mean annual egg production, 1899–1907 | 42 |
| | 10. Changes in annual egg production. Standard deviations, 1899–1907 | 45 |
| | 11. Changes in annual egg production. Coefficients of variation, 1899–1907 | 46 |
| | 12. Changes in annual egg production. Skewness, 1899–1907 | 47 |
| | 13. Changes in percentages of extreme variates in egg production, 1899–1907 | 48 |
| | 14. Mean annual egg production with different amounts of floor space per bird | 60 |
| | 15. Variability in annual egg production with different amounts of floor space per bird | 62 |
| | 16. Frequency polygons for egg production during the first and second years of laying | 64 |
| | 17. Change in intensity of selection, 1899–1907 | 70 |

7

# A BIOMETRICAL STUDY OF EGG PRODUCTION IN THE DOMESTIC FOWL.

## I. VARIATION IN ANNUAL EGG PRODUCTION.[a]

### INTRODUCTION.

In 1898 the Maine Agricultural Experiment Station began the use of trap nests[b] in connection with its poultry investigations. Since that time such nests have been used to obtain records of the egg production of all hens kept by the station. In consequence there has now accumulated a remarkable—in some respects, indeed, a unique—collection of exact statistical data regarding the phenomenon of egg production in certain breeds of the domestic fowl. So far as we are aware no other set of records of egg production exists which compares with this either in regard to the number of birds included in the statistics or in length of time covered by continuous daily records. On this account it has seemed desirable that these statistics be submitted to such thorough biometrical analysis as their intrinsic value obviously warrants. Hitherto practically no analytical study of egg production statistics (either the present collection or others) has ever been made. The need of such study, however, is apparent if it be considered that only by analysis of comprehensive statistics of egg production can we hope to reach a definite formulation of the fundamental biological laws underlying the process itself. The existing knowledge of egg production in fowls is on much the same plane as that regarding milk production in dairy cattle, for example. In both cases empirical methods have led to a knowledge of how to feed, care for, and, to some extent, breed the animals in order to get a high production of the thing desired. But of the biological factors

---

[a] Papers from the Biological Laboratory of the Maine Agricultural Experiment Station, No. 2.

[b] The particular type of trap nest used was designed at this station and has been described and figured in the Annual Report of the Maine Agricultural Experiment Station for 1898, pp. 141–143, in various bulletins of the station issued subsequently, and in Bulletin 90 of the Bureau of Animal Industry. Recently a new and improved trap nest has been devised and has replaced the earlier form of nest in the experimental work of the station. For a description of this new nest see Bulletin 159 of the Maine station.

concerned in improved egg production or milk secretion, and the laws according to which these factors operate, relatively little is definitely known.

In the case of the fowl the frequency of ovulation in a given period of time has been enormously increased in the domesticated hen as compared with the wild *Gallus bankiva*. It is commonly said that the great egg production of present-day "improved" varieties is due to the long-continued process of selection for this quality or characteristic which has been going on since domestication of fowls began. Possibly this is true in a broad sense, but it does not tell us anything about the underlying physiological factors which are the things that have been improved by this selection, or even what these factors are. Yet it is about them that we must know if we are to make advance in the direction of increasing egg production. Selection must be directed with ever-increasing precision as it goes on if improvement is to continue, and to make it more precise the breeder must have more detailed knowledge regarding the thing for which he is selecting, and how this is inherited.

A point of particular biological interest in connection with statistics of egg production lies in the relation which they bear to statistics of fecundity and fertility in other organisms. In nearly all animals other than birds the only possible measure of fecundity is based on the number of young produced. But the number of young produced depends not alone on the mother's inherent reproductive capacity, but also on her opportunities for getting the ova she produces fertilized. Statistics such as those with which we are here dealing are statistics of *ovulation*, and they make it possible to study this single phase of the problem of fecundity free from disturbing factors.[a]

The general purpose of this study is to make as thorough an analysis of the phenomenon of egg production in the domestic fowl as may be possible. The detailed statistics which we have available are so extensive and the specific problems to be attacked are so numerous that it has seemed advisable to publish the results of the investigation in separate parts. The general topics which it is tentatively proposed shall be discussed in the several parts are briefly indicated in the following outline:

I. Variation in egg production based on the study of the records of annual production. (The present paper.)

II. The egg production in the several months of the year and the relation of monthly to total annual egg production. This portion of the investigation has as its aim the "dissection" of the gross annual curves of variation in egg production.

---

[a] For a more detailed discussion of this point see Maine Agricultural Station Bulletin 166.

III. The effect of environmental influences on egg production.

IV. The relation of certain internal factors to egg production.

V. General summary and discussion of the results of the investigation.

The present paper (Part I) is in the first instance a necessary preliminary to those which are to follow. Its main purpose is, in a sense, to define the problems with which the subsequent parts of the work have to do. In it are set forth the facts regarding variation in first-year egg production.[a] These facts we shall attempt to analyze in later portions of the work. The subject of variation in first-year egg production has, however, a number of features of interest on its own account, and presents certain problems which can be answered without treating the more detailed statistics. First in importance is to learn definitely what the characteristics of variation in egg production in the individual are. To what kind of variation curves do statistics of ovulation lead? What are the biometrical constants of such curves? How does variation in annual egg production resemble or differ from variation in other characters, and in particular variation in fertility and fecundity? How are the variation constants for first-year production affected by environmental influences? In what way and to what extent has annual egg production been influenced by selective breeding? What relation exists between first-year and second-year production? These are some of the specific problems which will be considered in the paper.

### MATERIAL.

Before undertaking a detailed account of the group statistics it will be well to make some statement regarding the homogeneity—from the biological standpoint—of the material from which these records have been obtained. As has been stated in publications of the Maine Agricultural Experiment Station, there has been conducted at this station during the last nine years an investigation under the direction of the late Prof. Gilbert M. Gowell on the subject of breeding hens for increased egg production. The statistical material with which the present paper has to do comprises the egg records obtained in the course of this breeding investigation. It will not be necessary to state here in detail the plan of this breeding work, since that has already been thoroughly done in Bulletin 90 of the Bureau of Animal Industry. In that bulletin (pp. 7–15) is given a detailed account of the methods used in this breeding work up to the year 1906. A general statement regarding this work will suffice for our present purpose.

---

[a] The term "first-year egg production" will be used throughout as a convenient verbal contraction in the place of "the number of eggs produced by pullets in the year beginning the first day of the November following their hatching and ending a year from that date."

At the beginning of the experiment in 1898 three breeds of hens were used: (1) A strain of Barred Plymouth Rocks which had been bred by Professor Gowell for a period of some twenty-five years; (2) a strain of White Wyandottes which had been bred at the station for some years; and (3) a strain of Light Brahmas which similarly had been bred for some time at the station.

All the birds in hand at the beginning of the breeding experiment in 1898 were purebred in the ordinary sense and had for some years been bred by the same person—Professor Gowell. This insures that in regard to racial purity they form material which is, from a biological standpoint, very homogeneous. This is particularly true of the Barred Plymouth Rocks, which furnish the bulk of the statistical material with which the present investigation has to do. After October, 1902, no variety other than the Barred Plymouth Rock was used in the egg-production work. The White Wyandottes were used from 1898 to 1902, and egg records for them between these dates are available for study. On account of the small number of Light Brahmas available, together with the fact that they were dropped early in the course of the breeding experiment, they have not been included in the present investigation.

In breeding the birds from 1898 down to the present time the practice as stated in the published reports of the work has been to use each year, as mothers of the stock produced in that year, only hens which have, between November 1 of the year in which they were hatched and November 1 of the next following year, laid 160 or more eggs. The fathers of all stock raised since the breeding season of 1900 have been the sons of mothers whose production in their first laying year was 200 eggs or over. It will be seen again that this breeding practice has been such as to lead to great homogeneity of the statistical material from a biological standpoint. It is stated, however, that except in the first years of the breeding there has been no "close" inbreeding. No birds as closely related as first cousins are said to have been bred together since the first year of the breeding experiment.

The nature of the statistical records with which the present paper deals is briefly as follows: The first laying year of a hen has been arbitrarily taken to be from November 1 of the year in which she is hatched to November 1 of the following year. The "first year's egg record" of a hen, then, consists of the trap-nest record of the total number of eggs produced by that hen within such a year. Such "first-year" records constitute the basic material for Part I of this study.

Some important questions suggest themselves here. Has the "first-year" egg production any particular biological significance as compared with the egg production in any other (shorter or longer) period in the life of the bird? To what extent is there justification for taking first-year production as a measure of what might be called the total ability of a bird to produce eggs? In other words, why is annual egg production taken as the unit of the study of variation in this character? The answer to the last question is simplest: We are using annual production because it is the longest time unit for variation in egg production for which we have any extensive data available. A larger unit might, on some grounds, be desirable, but in such case as this the biometrician can only take the data as he finds them. It is obvious that the egg production in the first year can not be taken as a precise measure of the total number of eggs which hens would produce if allowed to live until natural death occurred. Whether it is a good measure or not depends on the degree to which first-year production is correlated with this unknown total possible production. No data exist from which this correlation can be precisely determined, but there are reasons to believe that if it could be evaluated it would be found to have a fairly high value. The chief of these reasons lies in the fact that in the domesticated fowl the average egg production of the first year is never equaled in any subsequent year, though with individual birds this relation may be changed. Thus, it is shown on page 65 that the mean production of 66 birds in their second laying year was but 61.7 per cent of that in their first year. That the average production falls off further in the third and subsequent years is known to be a fact. In other words, the first-year production, on the average, makes up a very considerably larger part of the unknown total production than does that of any other equal period of time. Hence we should expect it to be most highly correlated with that total.

The general conclusion which we have reached after careful consideration of the matter is that first-year production is to be regarded from the theoretical standpoint as a good measure of total possible production. It is a sufficiently large unit to include the influence of seasonal climatic changes, as well as those influences dependent on cyclical organic changes, such as, for example, original mating season, time of molting, and the like. From the practical standpoint first-year production is, for obvious reasons, in some respects the best unit which could be taken. It is highly desirable, however, that definite trap-nest records be obtained in considerable numbers for the egg production of the same birds during their entire lifetime or at least until they cease entirely to produce eggs. It is hoped that in time it will be possible to accumulate such records at the Maine station.

Another point to be noted is that the use of the term "first-year egg production" for the November-to-November laying is, in a sense, entirely arbitrary. Pullets very often begin to lay before

November 1 of the year in which they are hatched. Again, many pullets do not begin to lay until in December or later. They will not have laid a full year at the end of the following October. It is manifestly impossible to deal on any extensive scale with the actual first twelve months' egg production of each individual hen. One feasible procedure is to do as was done in this breeding investigation, namely, to take a calendar year as the unit of record, and choose the beginning of the year at a point which about averages the varying times of beginning egg production in the different individuals of the flock.

In extracting the egg records from the original record books certain complications arise which, for the sake of uniformity in the statistical treatment, we have had to decide in an arbitrary fashion. In the first place, it will always happen, of course, that death will prevent the completion of the "first laying year" by some of the birds which begin it. What shall be done with such records made incomplete by death? After careful study of the point it has been decided to omit entirely from our statistical treatment all records known to have been made incomplete by death. This is clearly the most logical method of dealing with such incomplete records which can consistently be applied. The same treatment has been given to the incomplete records of birds which were stolen during their first year. In the earlier years of the work it several times happened that birds were stolen after having completed only a part of the first year's record. All the records for stolen birds have been excluded from our statistics.

Another difficulty, and one in a sense more serious than either of those just mentioned, arises from the fact that hens do not invariably lay in the trap nests. It occasionally happens that birds will lay on the floor of the pen. Of course it then becomes impossible to connect the egg with the bird which laid it. When the study of these records was first undertaken it was for a time thought that this difficulty was a very serious one, and one which could not properly be allowed for in treating the data. Further study of the matter, however, led to the conclusion that this was not the case. In order that any conclusion regarding individual variation in egg production, or regarding the comparative production of flocks in different years or under different housing conditions, should be affected by the existence of unrecorded eggs laid on the floor, two conditions are necessary: (1) That the tendency to lay on the floor be not distributed at random (in the technical sense) among all the birds, and (2) that the total absolute amount of "floor" egg production should vary widely from year to year and in different pens. Are these conditions realized, as a matter of fact? We may discuss the two separately.

If it were a regular, constant, and undetected habit of some particular birds to lay outside the trap nests, and such birds formed a considerable portion of the flock, we could obviously draw no conclusions from trap-nest records. Such, however, appears not to be the case. Those who have had the birds in charge inform us that it has always been possible without any great difficulty to learn when a particular hen was beginning to form the habit of laying on the floor. Such hens when detected were trained, usually with comparative ease, to lay in the nests. A hen was never allowed to continue the floor-laying habit. All the available evidence and our own observations have convinced us that, could we know the birds that lay eggs outside the nests we should find them to be a random sample of the flock. That is to say, any bird is as likely as any other to lay an egg on the floor instead of in a trap nest. This being the case, the fact that there are eggs produced outside the nests will obviously not affect any conclusions respecting individual variation in egg production, if, as is the case, we deal with relatively large numbers of individuals.

Has the total number of eggs laid outside nests varied widely at different times during the period covered by the investigation, or has it been relatively constant? All the evidence that we have indicates that it has been relatively constant. Unfortunately, there exist no precise data on this point except for a single year (1904–5) out of the whole period covered by the statistics. We are informed by Mr. Walter Anderson, who has had immediate charge of the trap-nest records, that (a) the absolute amount of "floor" egg production has always been small, and (b) that in relation to the total number of eggs produced at any time the number laid on the floor has always been proportionately about the same. This means that figures for egg production based on the trap-nest records are absolutely a little too small in every case, but relatively are unaffected. It is relative figures with which we are chiefly concerned. It will obviously make no difference in our conclusions as to the comparative egg production of two years, say, if before making the comparison we add 2 per cent to the egg record for each year to cover the eggs laid on the floor.

The general conclusion, then, is that so far as the objects of the present investigation are concerned the unrecorded "floor" egg production is of no practical consequence. In the future poultry work of the Maine station an exact record will be kept of the eggs laid outside the nests.

It has further been decided to exclude from the statistical analysis the very few birds which, in the course of the trap-nesting experience of the station, have apparently laid no eggs at all in their first year. The number of such birds is so small (probably not 10 in the whole experience of the station) that it makes no significant difference in the statistical constants whether they are included or not. The specific reason for excluding such birds lies in an element of uncertainty in respect to the early records regarding them. That is, it is not possible now to be certain, in each of the cases where a bird has no record, whether the absence of record is due to the fact that she laid no eggs or to some other cause.

With this account of the general features of the statistics in hand, we may pass to the consideration of the records for particular years. We have decided to begin our discussion of the records with the data for the laying year 1899–1900; that is, with the records for birds hatched in April and May of the year 1899. The raw statistical data for this year have already been published.[a] They include records for the complete year (November 1, 1899, to November 1, 1900) for 70 Barred Plymouth Rocks and for the same number of White Wyandottes. An account of the housing and treatment of the birds for this year is given in the bulletin just referred to (pp. 31 and 32). So far as appears from the records, this laying year was in all respects a normal one. No serious accident or change of methods and management interfered in any way with the normal course of egg production.

The records for the laying year 1900–1901, that is, for birds hatched in April and May, 1900, have also been published.[b] Complete records for this laying year are available for 85 Barred Plymouth Rocks and 72 White Wyandottes. So far as may be judged from the records themselves and from the statements in Bulletin 79 (pp. 36 and 40) this may be considered to have been a normal laying year.

The egg records for the laying year 1901–2 have been published.[c] On account of lack of space relatively few pullets were carried over that winter. Complete records are available for but 48 Barred Plymouth Rocks and 33 White Wyandottes. The Barred Plymouth Rocks of that year were the first pullets sired by males bred from 200-egg-producing mothers to be tested in the trap nests. So far as we know this was a normal laying year.

The records for the years 1902–3 to 1906–7, inclusive, have never been published and were extracted by the authors of this paper from the original egg-record books in the archives of the station. In the year 1902–3, 147 of the Barred Plymouth Rock pullets tested in the trap nests completed first-year records. The egg production during this year can not be regarded as normal. The reason for this has been set forth by Professor Gowell[d] as follows:

The pullets were hatched in April and May, and thinking to have them mostly in readiness for laying early in November, we fed them rather more beef scrap than usual

[a] Maine Agricultural Experiment Station Bulletin 79, pp. 33–36. 1902.
[b] Loc. cit., pp. 37–40.
[c] Maine Agricultural Experiment Station Bulletin 93, pp. 72, 73. 1903.
[d] Maine Agricultural Experiment Station Bulletin 117, p. 96. 1905.

during the growing season, while they were out on the range, and before we were aware of their development they were laying—in August. They were nearly all laying heavily during September, October, and November. They were splendid birds, but almost every one of them molted completely in December, and we got very few eggs from them for more than two months. The most of the eggs secured from them were laid after the middle of January. Could they have commenced laying in October and continued for a year, molting would probably have been avoided and the showing would have been much better.

In the year 1903–4, 300 Barred Plymouth Rock pullets were tested in the trap nests. Of these 254 completed the first-year record and appear in the statistics of the present paper. The egg production of this year again can not be regarded as normal, for a reason which has been discussed by Professor Gowell (loc. cit.). Owing to a difficulty in the construction of a new house, the birds were not moved in from the range so as to begin their laying until December 6. Hence the records for 1903–4 as they appear in the statistics of the present paper are really eleven-month records instead of twelve-month records. They lack the potential contribution of the month of November, 1903, to the totals.

At the beginning of the laying year 1904–5 an experiment on the relation of the amount of floor space per bird to egg production was started. In carrying out this experiment the birds were divided into flocks of 50, 100, and 150 birds each. These were put into pens with floor space such that in the pens containing 50 and 100 birds there was 4.8 square feet of floor space per bird and in the 150-bird pens 3.2 square feet of floor space per bird. There were six 50-bird pens, and of the 300 birds which started the year in these pens 283 made complete records and appear in our statistics. There was one 100-bird pen, and 92 of these birds completed the year's record and appear in the present statistics. There was one 150-bird pen, and 140 of these birds completed the year's record and appear in the statistics. The records for this year are to be regarded as normal.

In the laying year of 1905–6 the birds tested in the trap nests were again divided into flocks of 50, 100, and 150 birds each. There were four of the 50-bird pens, and of these 200 birds 178 made complete records for the whole year and appear in the statistics of the present paper. There were two 100-bird pens, and 182 of these birds completed the year. There were also two 150-bird pens, and 275 out of the 300 birds completed the year. The laying year of 1905–6 was in nearly every respect a normal one. The only extraordinary factor which came in to interfere with the egg production during this year was a short period of sickness among the birds caused by the too free use of succulent feed. The details of this mishap have been stated by Professor Gowell [a] and need not be repeated here. The

a Maine Agricultural Experiment Station Bulletin 144, p. 184. 1907.

trouble occurred in April, 1906, and its effect on the egg yield is stated to have been as follows: "The egg yields were about 60 per cent of the number of birds just before the trouble began, but they were reduced to less than 10 per cent, and it was about twenty days before the birds regained their former production." This factor reduces the egg yields for the months of April and May of this year. The precise amount of this reduction would appear from the statistics not to be so great as is implied in the statement quoted. This will be more fully discussed in Part II of this study, which will deal particularly with monthly egg production.

In the laying year 1906–7 the experiments with flocks of different sizes were continued. We have records for this year of four 50-bird pens, two 100-bird pens, and two 150-bird pens. Of the 200 birds in the 50-bird pens 187 completed the year; of those in the 100-bird pens 185 completed the year; and, finally, of the 300 in the two 150-bird pens 281 have complete records for the year. In the early part of this laying year (December) a serious mishap occurred which considerably lowered the egg yields for this and the succeeding month. The matter is thus described by Professor Gowell:[a]

Last December the roosts and walls of the roosting closets in all of the rooms but one were sprayed with two different brands of liquid lice killers which were warranted by their makers to destroy all the lice on fowl by the fumes penetrating among the feathers to the skins of the birds. The preparations were used according to directions. The roosts and woodwork near them were sprayed in the morning and left to dry until the birds went to roost. But the sprayed paint was not fully dry at that time, although it was thinned by heating before applying it. The curtains of the roosting closets were shut down at bedtime as usual and both ventilators, each of which are 3 feet long and 6 inches wide, were left wide open. Next morning it was very evident from the appearance of the birds that they had not enjoyed the night. They ate but little food during that and the succeeding three or four days and did not have their usual appetites for nine or ten days. The 700 birds were laying over 300 eggs per day before the trouble, but they laid less than 100 per day during the following week, and did not lay as many as before until twenty-four days had elapsed.

Many of them molted partially, or quite fully, and these did not lay much for six or seven weeks. It is not thought that any birds died from the accident. Probably we were at fault in using the stuff in winter when it did not thoroughly dry out, as it might have done in a long, warm summer day. One pen at the end of the building was not sprayed, but the air in that room was loaded with the odor from the rest of the building, which easily found its way in around the loosely fitting door. The birds in that room fell off in their egg yields for several days, but none of them molted. Many of the birds in the sprayed rooms refused to go to roost in the closets again for several nights, until the odor had largely disappeared, while those in the unsprayed room went into their bedroom as usual.

The precise effect of this disturbance on the egg yields will be shown later in a discussion of the monthly production.

---

[a] Loc. cit., p. 185.

Another point which should be noted regarding the records for 1906–7 is that they comprise the records for eleven months instead of a full year. The records were discontinued at the end of September, 1907, October of that year being entirely omitted.

From the account of the statistics of the several years included in this investigation it is evident that as they stand the data for any one year are probably not strictly comparable with those for any other year. Certainly this is true in the later years covered by the material. The statistics for each year must be treated separately, and whenever comparisons between years are instituted great caution must be exercised in drawing conclusions. It will obviously be necessary in some cases to apply corrective factors before any comparisons whatever can be made. The authors are aware that this fact constitutes the principal point at which these statistics are open to criticism, and that it is a serious defect. It should be remembered, however, when such a criticism of data like these is made, that the practical difficulties of keeping relatively constant the environmental conditions under which so many as a thousand laying hens are kept are very great. Especially is this true when the time unit of the experiments is twelve months. In nearly every case the factors which have come in to disturb the records of annual egg production, as noted above, are things which could not have been foreseen and avoided. They were accidents in the strict sense. It will be apparent in the later portions of the paper that in spite of these accidents in the several years the constants of the variation curves of egg production are very closely similar throughout the period covered by the statistics. In other words, the effect of these accidents has been decidedly more on the centering constants like the mean and the mode than upon the constants which have to do with the shape of the curve, such as, for example, the standard deviation and the skewness. On the whole, we believe that the data which we have, in spite of their shortcomings, give us a very fair picture of the true state of affairs relative to variation in annual egg production. When we come to the detailed monthly and daily statistics, which will be considered in the parts of the bulletin following the present one, these factors, which to a certain extent vitiate annual statistics, will be very much less serious, because in every case the accident has been confined to but a single month or at most two or three months.

In discussing the variation in egg production we shall consider the statistics for the Plymouth Rocks first and base the major portion of the discussion upon them. The records for the White Wyandottes will be taken up farther on in a separate section of the paper.

## STATISTICAL METHODS USED.

In dealing with these egg-record statistics obviously the first thing to be done is to get the data in the form of frequency distributions. We have done this for each year's records of each set of pens (50, 100, and 150 birds) without grouping data in any way. These ungrouped frequency distributions, which form the basic raw material of the first part of this study, are given in detail in Appendix I, Tables I, II, and III. A glance at these distributions will indicate at once that the range of variation in egg production is very considerable. It is so great, in fact, as to make analytical treatment necessary if we are to form any adequate idea of variation in this character.

In this investigation we have made use of the ordinary methods of biometrical analysis. Summary accounts of these methods have been given by Elderton,[a] C. B. Davenport,[b] and E. Davenport.[c] For more detailed accounts the reader is referred to the memoirs of Pearson on the subject which have appeared in (a) the Philosophical Transactions of the Royal Society during the period from 1895 to 1905; (b) Biometrika; and (c) the Draper's Company Research Memoirs of the Department of Applied Mathematics of University College, London. In view of the existence of the treatises referred to it will not be necessary here to go into a general account of biometrical methods. It is, however, desirable to call attention to certain specific modes of treating these egg-production statistics which we have adopted.

The problem of grouping the material may be first considered. From the extent of the range it is obviously necessary that some grouping be resorted to in order to get a smooth frequency distribution covering the whole range of variation in annual egg production (more than 200 eggs). If the data were not grouped it would be necessary to have records of very many individuals, probably hundreds of thousands. With the smaller numbers which practical considerations make the only possible ones, it is clearly necessary that the data be grouped. When grouping such data the usual rule followed in biometrical work is to make as many classes as will give a fairly smooth distribution without too many classes having zero frequency. Obviously if the class units are sufficiently large all frequency distributions will be reduced to very much the same form. Only experience can decide what is the most appropriate value to take for the class unit in any specific case. After studying all the data on variation in egg production we have come to the conclu-

---

[a] Elderton, W. P. Frequency Curves and Correlation. London, 1907. Pp. xiii and 172.

[b] Davenport, C. B. Statistical Methods with Special Reference to Biological Variation. Second edition. New York, 1904. Pp. viii and 223.

[c] Davenport, E. The Principles of Breeding. Boston, 1907. Pp. xiii and 727.

sion that a class unit of 15 eggs is probably as satisfactory as any which could be taken.   In adopting this unit we have set the class limits at the same points in the case of each distribution; that is to say, the first class includes birds laying from 0 to 14 eggs inclusive; the next class, birds laying from 15 to 29 eggs inclusive, and so on. The procedure sometimes followed of taking different limits for the class unit in the case of each frequency distribution, and subsequently treating these distributions as if they were in every way strictly comparable, is obviously not a logical one.

In deducing the moments from the grouped material we have in each case used Sheppard's corrections.[a]   It will appear further on in the paper that the problem of getting satisfactory theoretical curves to graduate these statistics of egg production is an extremely difficult one.   It occurred to us that possibly better success in the graduation of the curves might be had by using raw instead of corrected moments.   Actual trial, however, showed that somewhat worse instead of better results were obtained when the raw moments were used.   On the whole, this is what is to be expected on theoretical grounds, as will be apparent when the grouped frequency distributions are examined.   Our material is grouped in rather large units, and at the same time we have apparently fairly close approximation to high contact at both ends of the range.   These conditions, of course, indicate that Sheppard's corrections are demanded.

In deducing constants from the ungrouped material we have calculated the mean and the median by ordinary arithmetical methods. In calculating standard deviations in these cases we have used the method of moments rather than the method of calculating from the formula $\sigma^2 = \dfrac{S(x^2)}{n}$ , where $x$ denotes deviation from the mean.   In calculating the mode in the case of the ungrouped frequency distributions, use has been made of the approximate relation:

Mean to mode $= 3 \times$ (mean to median),

the median of the curve lying between the mean and the mode. This approximation to the position of the mode was first worked out empirically by Pearson[b] and found to give reasonably close values in most cases.   Recently Charlier[c] has derived this relation directly in his treatment of a certain type of frequency curve. He shows definitely why this approximation to the mode must give a fairly close value in all cases where the frequency curve is not too widely divergent from the normal curve of errors.   In order to

a Sheppard, W. F.   Biometrika, Vol. V, pp. 450–459.   1907.

b Cf. Pearson, K.   Biometrika, Vol. I, pp. 260, 261.   1902.

c Charlier, C. V. L.   Researches into the Theory of Probability.   Meddelanden från Lunds Astronomiska Observatorium, Serie II, No. 4, Kongl.   Fysiografiska Sällskapets Handlingar, Bd. 16, pp. 1–51.   19 fig.   Lund, 1906.

determine for the class of curves with which we are here dealing how good this approximation to the mode is, we have calculated for six curves, (a) the true value of the mode from the moments, by the usual general formula,[a]

$$d = \tfrac{1}{2} \frac{\sqrt{\mu_2}\ \sqrt{\beta_1}\ (\beta_2 + 3)}{5\beta_2 - 6\beta_1 - 9}$$

and (b) the approximate value of the mode by the method under discussion.    The results are shown in the following table:

| Year and group. | True mode. | Mode by approximate method. | Difference. |
|---|---|---|---|
| 1903 | 139. 20 | 143. 25 | −4. 05 |
| 1904, 50-bird pens | 149. 14 | 148. 30 | −0. 84 |
| 1905, 50-bird pens | 149. 48 | 149. 88 | −0. 40 |
| 1905, 100-bird pens | 135. 00 | 133. 00 | +2. 00 |
| 1906, 50-bird pens | 123. 21 | 120. 44 | +2. 77 |
| 1906, 150-bird pens | 103. 73 | 105. 89 | −2. 16 |

When it is remembered that the probable errors of the centering constants (mean and median) are in these curves of about the magnitude ±2, it is apparent that the approximation gives values as close to the true mode as could reasonably be expected.

Having determined the mode in the manner indicated, in the case of the ungrouped material, the distance from mean to mode was used to find the skewness by the following relation:

$$\text{Skewness} = \frac{\text{Distance from mean to mode}}{\text{Standard deviation}}$$

Respecting the probable errors of the various constants discussed, the following should be said.    The probable error of the mean was calculated from the usual formula,

$$P.\ E._{mean} = 0.67449 \frac{\sigma}{\sqrt{n}}$$

using Gibson's tables.[b]    Theoretically, in the case of the grouped material, where Sheppard's corrections are applied to the moments, the uncorrected value of $\sigma$ should be used in this formula.[c]    But practically, with the very large values of $\sigma$ with which we have here to do, and the consequent large absolute values of the probable errors, it makes no difference, to the number of decimal places tabulated in the probable errors, whether the corrected or uncorrected

[a] Pearson, K.   Philosophical Transactions of the Royal Society, vol. 198 A, p. 277. 1902.

[b] Gibson, W.   Biometrika, Vol. IV, pp. 385–393.   1906.

[c] Cf. Sheppard, W. F.   Biometrika, Vol. V, p. 455.   1907.

value is used.   We have consequently used  the corrected values to save labor.   The probable error of the median is given by Sheppard[a] as follows:

$$P.\ E._{median} = 0.84535 \frac{\sigma}{\sqrt{n}}$$

For the sake of economy in calculation we have made use of the obviously simpler relation:

$$P.\ E._{median} = 1.25332 \times P.\ E._{mean}$$

In the case of the probable error of the standard deviation it has recently been pointed out by the writer[b]  that caution must be exercised in using the formula usually given, viz:

$$P.\ E.\ \sigma = 0.67449 \frac{\sigma}{\sqrt{2n}} \quad \text{-----------------} \quad (i)$$

When curves deviate much from the normal curve of errors $P.\ E.\ \sigma$ by this formula may be seriously in error.   Resort should be had in all cases of importance to the general formula

$$P.\ E.\ \sigma = 0.67449 \sqrt{\frac{\mu_4 - \mu_2^2}{4\mu_2} \Big/ n} \quad \text{------------} \quad (ii)$$

which for convenience in computation may be written

$$P.\ E.\ \sigma = \chi_1 \sqrt{\frac{\mu_4 - \mu_2^2}{4\mu_2}}$$

where $\chi_1$ is the value $0.6744898 / \sqrt{n}$ tabled by Miss Gibson (loc. cit.). In the present investigation we have made it a rule to calculate this probable error by equation (ii) in every case where it was necessary for some other reason to determine the value of $\mu_4$.   In all other cases equation (i) was used.   In the tables it will be specifically indicated which formula was used in each particular case.

The usual formula for the probable error of the coefficient of variation $V$,

$$P.\ E._v = \frac{.067449}{\sqrt{2n}} \left[ 1 + 2 \left( \frac{V}{100} \right)^2 \right]^{\frac{1}{2}}$$

was used throughout.

In the case of the skewness the question of the proper value of the probable error is a difficult one.   It has been shown by Pearson and Filon[c] that the probable error of this constant has a different value

---

a Sheppard, W. F.  Philosophical Transactions of the Royal Society, vol. 192 A, pp. 101–167.  1898.

b Pearl, R., Biometrika, Vol. VI, pp. 112–117.  1908.

c Pearson, K., and Filon, L. N. G.  Philosophical Transactions of the Royal Society; vol. 191 A, pp. 229–311.  1898.

for each different type of frequency curve.   A maximum limit, however, is always given by the value in the case of the normal curve, i. e.,

$$P.\,E._{sk.} = 0.67449\sqrt{\frac{3}{2n}}$$

The authors cited point out (loc. cit., p. 271, footnote) that a very close approximation of the probable error of the skewness generally is given by the relation

$$P.\,E._{sk.} = 0.67449\sqrt{\frac{3}{2n}}\,\frac{1}{\sqrt{\{1 + 3(Sk.)^2\}}}$$

This formula we have used throughout, first putting it in the somewhat simplified form

$$P.\,E._{sk.} = \frac{1.7320508\chi_2}{\sqrt{\{1 + 3(Sk.)^2\}}}$$

where $\chi_2$ is the function tabled by Miss Gibson (loc. cit.).

All the computations in this investigation have been carried out with the aid of a large Brunsviga arithmometer and an adding machine.   All possible precautions have been taken to exclude arithmetical errors in the work.   It is the intention to publish all the raw material which was available to the authors (in the present or later parts) of this paper, so that it will be possible for any interested persons to check up the computations.

### VARIATION IN EGG PRODUCTION IN BARRED PLYMOUTH ROCKS.

Turning now to the discussion of the results of the work, we have first the grouped frequency distributions for variation in first-year egg production in Barred Plymouth Rock pullets during a period of eight years set forth in Table 1.   In this table are given both the absolute frequencies and the frequencies per cent for each year.

Table 1.—*Frequency distributions for variation in first-year egg production, Barred Plymouth Rocks.*

| Year and group. | \| | Annual egg production. | | | | | | | | | | | | | | | | Total. |
|---|---|---|---|---|---|---|---|---|---|---|---|---|---|---|---|---|---|---|
| | 0-14. | 15-29. | 30-44. | 45-59. | 60-74. | 75-89. | 90-104. | 105-119. | 120-134. | 135-149. | 150-164. | 165-179. | 180-194. | 195-209. | 210-224. | 225-239. | 240-254. | |
| **1899-1900:** | | | | | | | | | | | | | | | | | | |
| Absolute frequency | | | 3 | 2 | 4 | 4 | 4 | 2 | 10 | 11 | 9 | 12 | 6 | 1 | 1 | 1 | | 70 |
| Percentage frequency | | | 4.29 | 2.86 | 5.71 | 5.71 | 5.71 | 2.86 | 14.29 | 15.71 | 12.86 | 17.14 | 8.57 | 1.43 | 1.43 | 1.43 | | 100.00 |
| **1900-1901:** | | | | | | | | | | | | | | | | | | |
| Absolute frequency | 1 | | | 4 | 4 | 2 | 6 | 7 | 6 | 15 | 5 | 14 | 12 | 7 | 1 | 1 | | 85 |
| Percentage frequency | 1.18 | | | 4.71 | 4.71 | 2.35 | 7.06 | 8.24 | 7.06 | 17.65 | 5.88 | 16.47 | 14.12 | 8.24 | 1.18 | 1.18 | | 100.03 |
| **1901-1902:** | | | | | | | | | | | | | | | | | | |
| Absolute frequency | | | | | 1 | 1 | 2 | 8 | 4 | 1 | 11 | 8 | 3 | 5 | 3 | | 1 | 48 |
| Percentage frequency | | | | | 2.08 | 2.08 | 4.16 | 16.67 | 8.32 | 2.08 | 22.92 | 16.67 | 6.25 | 10.42 | 6.25 | | 2.08 | 99.98 |
| **1902-1903:** | | | | | | | | | | | | | | | | | | |
| Absolute frequency | | 2 | | 1 | 5 | 8 | 17 | 18 | 17 | 26 | 17 | 18 | 9 | 2 | 6 | 1 | | 147 |
| Percentage frequency | | 1.36 | | .68 | 3.40 | 5.44 | 11.56 | 12.24 | 11.56 | 17.69 | 11.56 | 12.24 | 6.12 | 1.36 | 4.08 | .68 | | 99.97 |
| **1903-1904:** | | | | | | | | | | | | | | | | | | |
| Absolute frequency | 7 | 5 | 5 | 10 | 10 | 20 | 24 | 29 | 52 | 37 | 29 | 16 | 8 | 2 | | | | 254 |
| Percentage frequency | 2.76 | 1.97 | 1.97 | 3.94 | 3.94 | 7.87 | 9.45 | 11.42 | 20.47 | 14.57 | 11.42 | 6.29 | 3.15 | .79 | | | | 100.01 |
| **1904-1905:** | | | | | | | | | | | | | | | | | | |
| *50-bird pens—* | | | | | | | | | | | | | | | | | | |
| Absolute frequency | 3 | 3 | 14 | 7 | 9 | 15 | 16 | 39 | 24 | 40 | 33 | 28 | 16 | 25 | 7 | 3 | 1 | 283 |
| Percentage frequency | 1.06 | 1.06 | 4.95 | 2.47 | 3.18 | 5.30 | 5.65 | 13.78 | 8.48 | 14.13 | 11.66 | 9.89 | 5.65 | 8.83 | 2.47 | 1.06 | .35 | 99.97 |
| *100-bird pens—* | | | | | | | | | | | | | | | | | | |
| Absolute frequency | | 2 | 4 | 1 | 3 | 8 | 10 | 11 | 5 | 7 | 11 | 11 | 8 | 9 | 2 | | | 92 |
| Percentage frequency | | 2.17 | 4.35 | 1.09 | 3.26 | 8.70 | 10.87 | 11.96 | 5.43 | 7.61 | 11.96 | 11.96 | 8.70 | 9.78 | 2.16 | | | 100.01 |
| *150-bird pens—* | | | | | | | | | | | | | | | | | | |
| Absolute frequency | 7 | 6 | 3 | 9 | 6 | 15 | 10 | 18 | 12 | 10 | 19 | 11 | 6 | 6 | 2 | | | 140 |
| Percentage frequency | 5.00 | 4.29 | 2.14 | 6.43 | 4.29 | 10.71 | 7.14 | 12.86 | 8.57 | 7.14 | 13.57 | 7.86 | 4.29 | 4.29 | 1.43 | | | 100.01 |
| **1905-1906:** | | | | | | | | | | | | | | | | | | |
| *50-bird pens—* | | | | | | | | | | | | | | | | | | |
| Absolute frequency | | | 1 | 2 | 4 | 9 | 13 | 25 | 24 | 22 | 32 | 17 | 20 | 9 | | | | 178 |
| Percentage frequency | | | .56 | 1.12 | 2.25 | 5.06 | 7.30 | 14.04 | 13.48 | 12.36 | 17.98 | 9.55 | 11.24 | 5.06 | | | | 100.00 |
| *100-bird pens—* | | | | | | | | | | | | | | | | | | |
| Absolute frequency | 1 | 3 | 3 | 3 | 8 | 14 | 17 | 26 | 25 | 23 | 27 | 16 | 12 | 3 | | | 1 | 182 |
| Percentage frequency | .55 | 1.65 | 1.65 | 1.65 | 4.40 | 7.69 | 9.34 | 14.29 | 13.74 | 12.64 | 14.84 | 8.79 | 6.59 | 1.65 | | | .55 | 100.02 |
| *150-bird pens—* | | | | | | | | | | | | | | | | | | |
| Absolute frequency | 1 | 1 | 4 | 10 | 21 | 23 | 35 | 46 | 40 | 35 | 25 | 19 | 8 | 6 | 1 | | | 275 |
| Percentage frequency | .36 | .36 | 1.45 | 3.64 | 7.64 | 8.36 | 12.73 | 16.73 | 14.55 | 12.73 | 9.09 | 6.91 | 2.91 | 2.18 | .36 | | | 100.00 |
| **1906-1907:** | | | | | | | | | | | | | | | | | | |
| *50-bird pens—* | | | | | | | | | | | | | | | | | | |
| Absolute frequency | 2 | 2 | 5 | 5 | 9 | 16 | 30 | 39 | 26 | 21 | 19 | 12 | 1 | | | | | 187 |
| Percentage frequency | 1.07 | 1.07 | 2.67 | 2.67 | 4.81 | 8.56 | 16.04 | 20.86 | 13.90 | 11.23 | 10.16 | 6.42 | .53 | | | | | 99.99 |
| *100-bird pens—* | | | | | | | | | | | | | | | | | | |
| Absolute frequency | 4 | | 8 | 11 | 7 | 25 | 29 | 25 | 30 | 20 | 11 | 10 | 5 | | | | | 185 |
| Percentage frequency | 2.16 | | 4.32 | 5.95 | 3.78 | 13.51 | 15.68 | 13.51 | 16.22 | 10.81 | 5.95 | 5.41 | 2.70 | | | | | 100.00 |
| *150-bird pens—* | | | | | | | | | | | | | | | | | | |
| Absolute frequency | 10 | 8 | 8 | 15 | 29 | 32 | 48 | 39 | 36 | 25 | 18 | 6 | 5 | | 2 | | | 281 |
| Percentage frequency | 3.56 | 2.85 | 2.85 | 5.34 | 10.32 | 11.39 | 17.08 | 13.88 | 12.81 | 8.90 | 6.40 | 2.14 | 1.78 | | .71 | | | 100.01 |

The percentage distributions for 1903, for the 50-bird pens of 1904, the 50, 100, and 150 bird pens of 1905, and the 50, 100, and 150 bird pens of 1906 are shown graphically in figures 1 to 8, inclusive. Of the other distributions the majority involve so few individuals relatively that they give rather irregular polygons. On this account, and since the general form in all is very similar to that of the polygons which are shown graphically, it will not be necessary to figure them separately.

From the data given in Table 1 and the diagrams we note at once the following general characteristics of variation in egg production:

1. The distribution of variation is not usually symmetrical about the center or type of egg production. The distributions nearly all show some degree of asymmetry or skewness. Further, this skewness is obviously negative in all cases—that is, the distributions tail off more gradually toward the side of low egg production.

2. The range of variation in egg production is very considerable, even in a stock so closely selected to type as this has been. The observed range extends from an egg production of zero to one of nearly 250. The manner in which this great range is covered is perhaps even more strikingly shown in the tables of Appendix I than in Table 1.

3. The distributions generally have more than one observational mode or peak. This fact, however, is probably not to be interpreted as meaning that the normal egg-production variation curve is a multimodal one. It is in all probability merely a result of random sampling with respect to a character having a great range of variation, when, furthermore, each sample contains only a relatively small number of individuals. The general form of the frequency distributions would indicate that the variation in the character is of the usual unimodal skew type. This point, however, will be more particularly discussed later on in the paper, when the evidence from the analytical constants is at hand.

4. Not only is the range of variation in egg production great, but also the absolute amount of variation in the character, as indicated by the "scatter" of the distributions, is relatively large. This will be made clearer by the constants of variation, to the examination of which we may next turn.

In Table 2 are given the values of the chief physical constants of the frequency distributions of Table 1, together with their probable errors. The methods by which the calculations were made have been described in the previous section. Six constants are included in the table. A brief examination of their meaning may be useful. The first of the constants tabled—the mean or average annual egg production—needs no explanation. The median (the second constant tabled) is the value at which the flocks divide into two equal parts in regard to egg production. Thus, in 1902–3 one-half of the birds laid more than 137.5 eggs and one-half laid fewer than that number. The mode is that amount of annual egg production which is exhibited by the largest number of birds. It is the "fashionable" egg production. For example, in 1902–3 more hens laid 139.53 eggs in the year than laid any other single number, either greater or less. The standard deviation is a measure of absolute variation; in the present instance it expresses the degree of variation in annual egg production in terms of eggs. It is a measure of the "spread" or "scatter" of the frequency distribution. The coefficient of variation is a relative measure of variation. It is the percentage of the standard deviation in the mean. The skewness is the measure of the degree to which the frequency curve is asymmetrical. A positive skewness means that the curve tails off more gradually toward large values than toward small values of the character under discussion, while a negative skewness means that, as in the present cases, the curve tails off more gradually toward the small values of the character. With positive skewness the mean is greater than the mode, while with negative skewness the reverse relation holds.

In the last three lines of Table 2 are given the weighted mean values of each constant for the years 1904–5, 1905–6, and 1906–7, respectively. In each of these years we have data for flocks of 50, 100, and 150 birds each. In order to get a single value which shall be in a measure representative of each constant for each of these years the only feasible procedure is to take a weighted mean of the values of the particular constants for each of the three flocks. This we have done. We have weighted the several flock constants in each case with the number of birds they involve, as shown in Table 1. A single example worked out in extenso will make plain just how these weighted values are obtained:

$$1905\text{–}6, \ 50\text{-bird pens: Mean } 140.31 \times 178 = \ 24{,}975.18$$
$$1905\text{–}6, \ 100\text{-bird pens: Mean } 127.50 \times 182 = \ 23{,}205.00$$
$$1905\text{–}6, \ 150\text{-bird pens: Mean } 119.43 \times 275 = \ 32{,}843.25$$

$$\text{Sums} = 635 \text{ and } 81{,}023.43$$

$81{,}023.43 \div 635 = 127.60 =$ "weighted mean" egg production in year 1905–6.

TABLE 2.—*Constants of variation in annual egg production, Barred Plymouth Rocks.*

| Laying year and group. | Mean.[a] | Median. | Mode. | Standard deviation. | Coefficient of variation. | Skewness. |
|---|---|---|---|---|---|---|
| 1899–1900............ | 136.36±3.55 | 142.00±4.45 | 153.29 | 44.03±2.51 | 32.29±2.02 | −0.384±0.082 |
| 1900–1901............ | 143.44±3.29 | 146.25±4.12 | 151.88 | 45.02±2.33 | 31.38±1.78 | −.188±.081 |
| 1901–2.............. | 155.58±3.79 | 158.50±4.75 | 164.33 | 38.94±2.68 | 25.03±1.83 | −.225±.104 |
| 1902–3.............. | 136.48±2.19 | 137.50±2.74 | 139.54 | 39.34±1.55 | 28.82±1.22 | −.078±.067 |
| 1903–4.............. | 117.87±1.75 | 126.33±2.20 | 139.20 | c41.43±1.31 | 35.14±1.17 | −.515±.039 |
| 1904–5: | | | | | | |
| 50-bird pens....... | 134.60±1.98 | 139.17±2.49 | 149.14 | c49.49±1.29 | 36.77±1.17 | −.294±.044 |
| 100-bird pens...... | 133.61±3.47 | 141.00±4.35 | 155.78 | c49.41±2.46 | 36.98±2.07 | −.449±.068 |
| 150-bird pens...... | 114.54±3.02 | 117.00±3.79 | 121.93 | c53.04±2.38 | 46.31±2.23 | −.139±.068 |
| 1905–6: | | | | | | |
| 50-bird pens....... | 140.31±1.81 | 143.50±2.26 | 149.48 | c35.75±1.11 | 25.48±0.97 | −.256±.052 |
| 100-bird pens...... | 127.50±2.03 | 129.33±2.54 | 135.00 | c40.59±1.49 | 31.84±1.23 | −.185±.056 |
| 150-bird pens...... | 119.43±1.54 | b118.86±1.93 | 120.07 | c37.88±1.01 | 31.72±1.00 | −.017±.050 |
| 1906–7: | | | | | | |
| 50-bird pens....... | 114.16±1.74 | 116.25±2.18 | 123.21 | c35.22±1.29 | 30.91±1.18 | −.263±.055 |
| 100-bird pens...... | 108.53±1.92 | b110.75±2.41 | 109.55 | c38.71±1.31 | 35.54±1.39 | −.016±.061 |
| 150-bird pens...... | 101.08±1.64 | 102.50±2.06 | 103.73 | c40.87±1.16 | 40.43±1.32 | −.065±.050 |
| 1904–5, weighted mean | 128.97 | 133.47 | 142.93 | 50.44 | 39.40 | −.280 |
| 1905–6, weighted mean | 127.60 | 128.77 | 132.59 | 38.06 | 30.01 | −.132 |
| 1906–7, weighted mean | 106.94 | 108.77 | 110.96 | 38.64 | 36.32 | −.108 |

[a] Beginning with the year 1902–3, the means (and also the other constants except the medians) are calculated from the grouped frequency distributions of Table 1, instead of from the ungrouped distributions given in the appendix. This leads to slightly different values than are obtained if one calculates directly from the ungrouped material. The differences, however, are always very small, and are not sensible in comparison with the probable errors of the constants. In passing it should be said that on theoretical grounds there is every reason to suppose that the means from the grouped figures represent more nearly the unknowable true means than do those from the ungrouped material.

[b] The failure of the median in these cases to fall between the mean and the mode arises from the fact that the skewness of these curves is very small. Consequently mean, median, and mode very nearly coincide. Each is subject to a considerable probable error, great enough indeed so that twice the range included within the probable error of any one of these constants includes the other two within its limits.

[c] These probable errors were calculated by the formula involving $\mu_4$. (Cf. p. 23, supra.)

Attention may first be called to certain general points brought out by the constants in this table, and then turned to the discussion of specific problems, for which the table furnishes data. We note first that the mean or average egg production ranges in amount between about 100 eggs (150-bird pen, 1906–7) and 156 eggs (1901–2). The median and the mode are in every case larger than the mean, indicating, as does also the last column of the table, that all these egg-production variation curves are negatively skew. The amount of variation in annual egg production is large both absolutely and relatively. The lowest coefficient of variation in the table is 25.03 per cent (1901–2). Even this is a distinctly large coefficient, when compared with what has been found for various characters in other organisms than the domesticated fowl. The values of the constants measuring variation obtained in this case indicate that egg production is to be classed, so far as variation is concerned, with purely physiological characters. It has been pointed out by the writer[a] that in the case of man characters which in general may be regarded as strictly physiological (even though they may be, as in the present case, measured in morphological units) vary much more than do such morphological characteristics as bone measurements and the like. This point has recently been confirmed by Kellicott[b] in his study

[a] Pearl, R. Biometrika, Vol. IV, pp. 31–36. 1905.

[b] Kellicott, W. E. Journal of Experimental Zoology, Vol. IV, pp. 576–614. 1907.

of variation in the toad. We hope later to have quantitative data regarding variation in other characters of the domestic fowl, so that direct comparison of various characters and organs in respect to variation may be made.

It is of interest to compare these results on variation in egg production with the results of other workers regarding variation in fertility and fecundity in other organisms. In Table 3 we have collected the existing data, from which such comparisons may be made. We have included data on variation in number of seeds produced in a plant, this being obviously a measure of gross fertility in such a case.

TABLE 3.—*Constants of variation in fertility and fecundity in various organisms.*

| Subject. | Character. | Coefficient of variation. | Skewness. | Authority. |
|---|---|---|---|---|
| *Nelumbium luteum*................... | Number of seeds...... | 17. 445 | +0. 0164 | Pearl.[a] |
| Poland-China swine.................... | Size of litter........... | 27. 411 | + .0701 | Surface.[b] |
| Duroc-Jersey swine.................... | .....do.............. | 25. 997 | + .0530 | Surface.[b] |
| Horse................................ | Fecundity [c] .......... | 24. 771 | − .1286 | Pearson.[d] |
| Man................................. | Number of children... | 48. 41 | + .0802 | Powys.[e] |
| Domestic fowl........................ | Annual egg production | [f] 34. 21 | [f] − .205 | This paper. |

[a] American Naturalist, Vol. XL, pp. 757-768. 1906.
[b] Reductions of data given by Rommel (Bureau of Animal Industry Circular No. 95). A paper giving a detailed account of these reductions is now in press.
[c] Fecundity in this case means the fraction which the actual number of offspring arising from a given number of coverings is of the possible number of offspring under the circumstances.
[d] Biometrika, Vol. I, pp. 289-292. 1902. Actually only the moments of this fecundity curve are given at the place cited. From the moments we have calculated the coefficients of variation and the skewness.
[e] Biometrika, Vol. V., p. 251. 1905.
[f] Weighted mean value of the constant for the whole period 1899-1900 to 1906-7, inclusive. The weighting is in proportion to the number of birds involved in each year.

Further data for man might be included in this table, but that is hardly necessary, as the example given of a human fertility curve is based on very extensive data and may be regarded as typical. It is clear from the table that these egg-production curves are in no way exceptional in regard to the degree of variation which they exhibit. It appears to be generally the case that fecundity, fertility, and closely allied characters are highly variable. There would seem to be a tendency for the coefficients measuring variation in these characters to be above a value of 20 per cent. In general it may be said that, so far as indicated by the present data, the domestic fowl exhibits a relative degree of variation in annual egg production distinctly less than that shown in normal human stocks in respect to size of family, and rather greater than that in Thoroughbred brood mares in respect to fecundity.

In Table 3 have also been included comparative data respecting the skewness of variation in the several characters enumerated. It is evident at once that there is even less uniformity in this particular than in the relative amount of variation. Even the sign of the skewness is not the same in all cases. In general the degree of skewness is small. The only cases in which the direction of the asymmetry of the distribution is negative are those of fecundity in brood mares and egg production in fowls.    Human fertility curves all agree in showing a distinct positive skewness.    It was at first thought that the negative skewness of the egg-production curves might owe its origin to the practice which has been followed of selection for high production.    It seems improbable, though, that this is the case, because, as indicated in the last column of Table 2, environmental conditions appear to have a great deal to do with the shape of the curves of variation in egg production.

### ANALYTICAL DISCUSSION OF VARIATION IN EGG PRODUCTION.

Having now considered the general facts respecting variation in annual egg production, we may next turn our attention to the more complete mathematical analysis of the variation curves, using the methods of Pearson.[a]    We have carried out the analysis in detail and fitted curves to the following distributions given in Table 1: 1903–4; 1904–5, 50-bird pens; 1905–6, 50, 100, and 150 bird pens; 1906–7, 50, 100, and 150 bird pens.    The other distributions involve so few individuals as to make their detailed treatment unprofitable.    The analytical constants are exhibited in Table 4.    The moments are given in terms of the unit of grouping, which in each case is equal to 15 eggs.

TABLE 4.—*Analytical constants of variation in annual egg production.*

| Constant. | 1903–4. | 1904–5, 50-bird pens. | 1905–6. | | | 1906–7. | | |
|---|---|---|---|---|---|---|---|---|
| | | | 50-bird pens. | 100-bird pens. | 150-bird pens. | 50-bird pens. | 100-bird pens. | 150-bird pens. |
| $N$ | 254 | 283 | 178 | 182 | 275 | 187 | 185 | 281 |
| $\mu_2$ | 7.6269 | 10.8868 | 5.6818 | 7.3233 | 6.3761 | 5.5117 | 6.6601 | 7.4221 |
| $\mu_3$ | − 14.6883 | − 13.3924 | − 3.8863 | − 7.0879 | .4478 | − 6.2369 | − 4.5912 | − 3.0085 |
| $\mu_4$ | 187.6675 | 318.7131 | 81.2274 | 168.5017 | 111.2316 | 97.7818 | 126.9814 | 165.2723 |
| $\beta_1$ | .4860 | .1390 | .0623 | .1279 | .0006 | .2323 | .0714 | .0221 |
| $\sqrt{\beta_1}$ | .6971 | .3728 | .2869 | .3577 | .0278 | .4820 | .2671 | .1488 |
| $\beta_2$ | 3.2262 | 2.6890 | 2.5161 | 3.1419 | 2.7360 | 3.2188 | 2.8627 | 3.0002 |
| $\beta_2-3$ | + .2262 | − .3100 | − .4839 | + .1419 | − .2640 | + .2188 | − .1373 | + .0002 |
| $\kappa$ | − 1.0055 | − 1.0389 | − 1.2149 | − .0999 | − .5303 | − .2594 | − .4886 | − .0660 |
| Skewness | − .5148 | − .2937 | − .2564 | − .1848 | − .0171 | − .2631 | − .0160 | − .0650 |
| $d$...........Eggs.. | 21.33 | 14.53 | 9.17 | 7.50 | .65 | 9.26 | .62 | 2.66 |
| Standard deviation, eggs. | 41.43 | 49.49 | 35.75 | 40.59 | 37.89 | 35.22 | 38.71 | 40.87 |
| Mean.........Eggs.. | 117.87 | 134.60 | 140.31 | 127.50 | 119.43 | 113.94 | 108.93 | 101.08 |
| Mode.........do... | 139.20 | 149.14 | 149.48 | 135.00 | 120.07 | 123.21 | 109.55 | 103.73 |
| Range.........do... | 331.82 | 328.21 | 221.29 | 1265.08 | 344.88 | 631.22 | 599.98 | .......... |
| +end of range..do.... | 194.41 | 248.18 | 208.55 | 313.77 | 291.87 | 226.11 | 358.93 | .......... |
| −end of range..do.... | −137.41 | − 80.02 | 12.74 | −951.31 | − 53.01 | −405.11 | −141.06 | .......... |
| $V$ per cent | 13.47 | 11.46 | 15.32 | 14.86 | 14.49 | 17.19 | 15.18 | 14.64 |
| P. E. $\sqrt{\beta_1}$[b] | ± .1037 | ± .0982 | ± .1238 | ± .1224 | ± .0996 | ± .1208 | ± .1214 | ± .0986 |
| P. E. $\beta_2$[b] | ± .2074 | ± .1964 | ± .2476 | ± .2448 | ± .1992 | ± .2416 | ± .2428 | ± .1972 |
| P. E. skewness[b] | ± .0518 | ± .0491 | ± .0619 | ± .0612 | ± .0298 | ± .0604 | ± .0607 | ± .0493 |

[a] Mathematical Contributions to the Theory of Evolution. II.—Skew Variation in Homogeneous Material. Philosophical Transactions of the Royal Society, vol. 186 A, pp. 343–414.  1895.  Math. Contr., etc.  X.—Supplement to a Memoir on Skew Variation.  *Ibid.*, vol. 197 A, pp. 443–459.  1901.
[b] These are the probable errors in the case of the normal curve, and are introduced to show the degree to which the present curves deviate from normality.

Examining first the criterion $\kappa_1 = 2\beta_2 - 3\beta_1 - 6$, we note that it is in every case negative. This indicates that unless $\sqrt{\beta_1} = 0$ within the limits imposed by its probable error the curves demanded are of Pearson's Type I, with range limited in both directions, and skewness. From the value of $\sqrt{\beta_1}$, given in the table, we conclude that the first four of the distributions, and also that for the 50-bird pens in 1906–7, fulfill the conditions for a Type I curve. The three remaining distributions have a very small (practically insignificant) skewness. Consequently $\sqrt{\beta_1}$ approaches zero closely in its value in these cases, and is in each instance not significant in comparison with its probable error. Hence, the conclusion is that a symmetrical curve must be used to graduate these three distributions. This limits the choice to Pearson's Type II, a symmetrical curve with the range limited at both ends, on the one hand, or to the normal or Gaussian curve of errors, symmetrical but of unlimited range, on the other hand. The criteria which indicate the proper choice in such a case are, first, that if $\kappa_1$ be negative (as is the case with all of our three curves) it indicates a Type II curve, and second, that if $\beta_2$ be significantly different from 3, again a Type II curve is indicated. In our three distributions $\beta_2 - 3$ can not in any single case be asserted to be significant in comparison with the probable error of $\beta_2$. In the case of the 1905–6 150-bird pen distribution the probability is that $\beta_2 - 3$ is significant. There is little doubt that in this case we shall get a better graduation with a Type II curve than with a normal curve. Accordingly, such a curve has been fitted in this distribution. In the case of the other two distributions we should undoubtedly get sensibly as good graduations with either Type II or normal curves. We have used Type II in the case of the 100-bird pen curve (1906–7) and the normal curve in the case of the 150-bird pen distribution (1906–7).

The equations to the eight curves follow.

1903–4:

$$y = 34.22 \left(1 + \frac{x}{18.4410}\right)^{6.9897} \left(1 - \frac{x}{3.6804}\right)^{1.3950}$$

1904–5, 50-bird pens:

$$y = 32.42 \left(1 + \frac{x}{15.2774}\right)^{4.8540} \left(1 - \frac{x}{6.6032}\right)^{2.0980}$$

1905–6, 50-bird pens:

$$y = 27.42 \left(1 + \frac{x}{9.1157}\right)^{3.3313} \left(1 - \frac{x}{4.7876}\right)^{1.7496}$$

1905–6, 100-bird pens:

$$y = 27.05 \left(1 + \frac{x}{72.4209}\right)^{108.1468} \left(1 - \frac{x}{11.9179}\right)^{16.8097}$$

1905–6, 150-bird pens:

$$y = 39.86 \left(1 - \frac{x^2}{132.1586}\right)^{8.8636}$$

1906–7, 50-bird pens:

$$y = 32.14 \left(1 + \frac{x}{35.2214}\right)^{86.7812} \left(1 - \frac{x}{6.8600}\right)^{7.1638}$$

1906–7, 100-bird pens:

$$y = 28.08 \left(1 - \frac{x^2}{277.7664}\right)^{19.3530}$$

1906–7, 150-bird pens:

$$y = 41.15\, e^{-0.0674x^2}$$

The frequency distributions and their fitted curves are shown graphically in figures 1 to 8, inclusive. In drawing these variation curves the frequencies have all been reduced to percentages. Hence, since the abscissal units are taken equal, all the curves have equal areas, and any given ordinates are directly comparable in the different diagrams.

It is clear from these figures that the curves give graduations of the data which are on the whole very good. They are quite as satisfactory as could reasonably be expected with such relatively small numbers in the observational series and the consequent irregularity of the polygons.

From Table 4 and the diagrams we note—

1. That the skewness or asymmetry of the variation curves, while in all but three cases distinct and significant, is in none of these curves extreme. Variation in egg production tends to approach the Gaussian type of distribution rather closely. The uniform tendency for the skewness in these curves to be in the negative direction when it exists at all is evident from the diagrams.

2. There is no uniform tendency for these ovulation curves to be either leptokurtic or platykurtic (that is, more or less peaked than the normal curve of equal standard deviation). In four out of the eight curves the sign of $\beta_2 - 3$ is positive, in the remainder negative. The deviation from the condition of mesokurtosis is in no case great, however. In view of the magnitude of the probable errors of $\beta_2$, it is impossible to assert that any of the deviations from mesokurtosis are certainly significant. In the case of the 1905–6 50-bird pen curve the value of the kurtosis $\beta_2 - 3$ is probably significant. In all the other cases the probability is in the other direction.

3. The total range is clearly overestimated in all but one case (1905–6, 50-bird pen), and in several of the curves this exaggeration of the range is very great. A total range of 328 or 332 or 345, while rather large, is not absolutely impossible theoretically. But when we come to a range of 1,265 eggs for the individual variation in egg

production within a year, it is clear that the exaggeration has gone far beyond the limits of what we must regard as possible. In spite

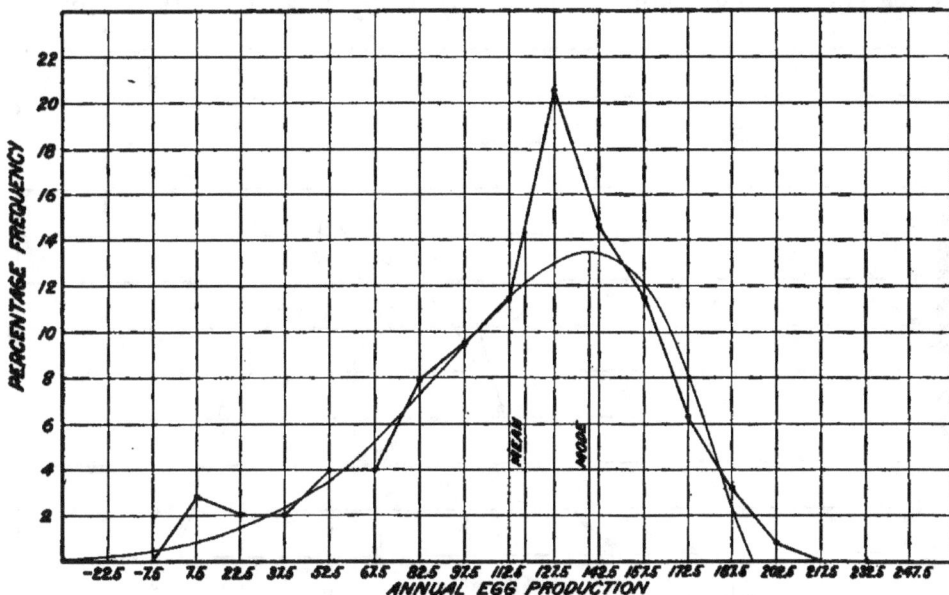

FIG. 1.—Variation in annual egg production, 1903–4.

of the great overestimation of the total range it will be observed that the upper (+) end of the range in every case has a value which is theoretically possible. While the production of over 300 eggs in a

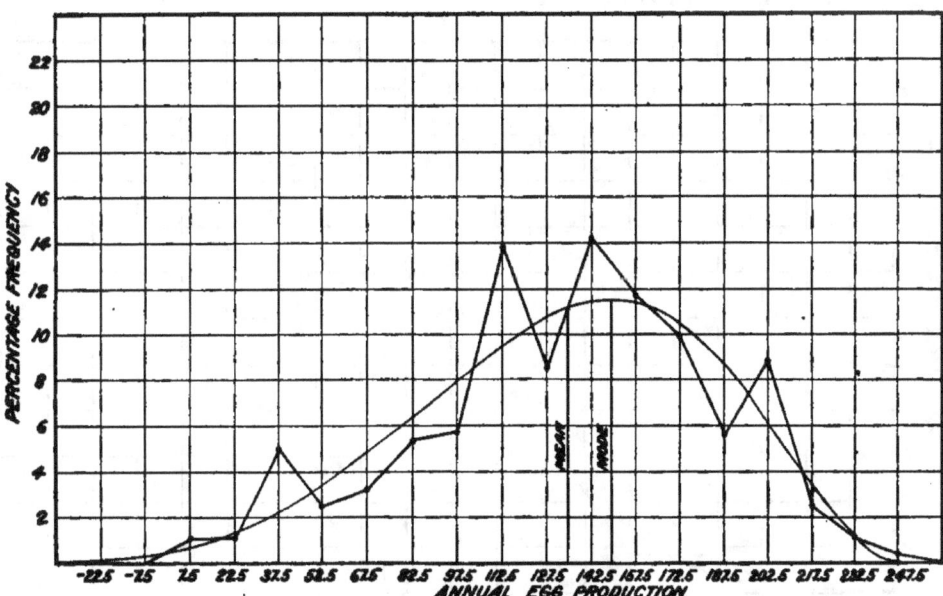

FIG. 2.—Variation in annual egg production, 1904–5.

year by an individual hen has never been recorded so far as we are aware, it still would be very difficult to prove that such an event is theoretically impossible. The really serious difficulty with the range, as given by the curves, lies in the location of the lower (−) range end. In only one of the eight curves (1905–6, 50-bird pen) does the lower range end have a value which is physically possible.   It is obvious

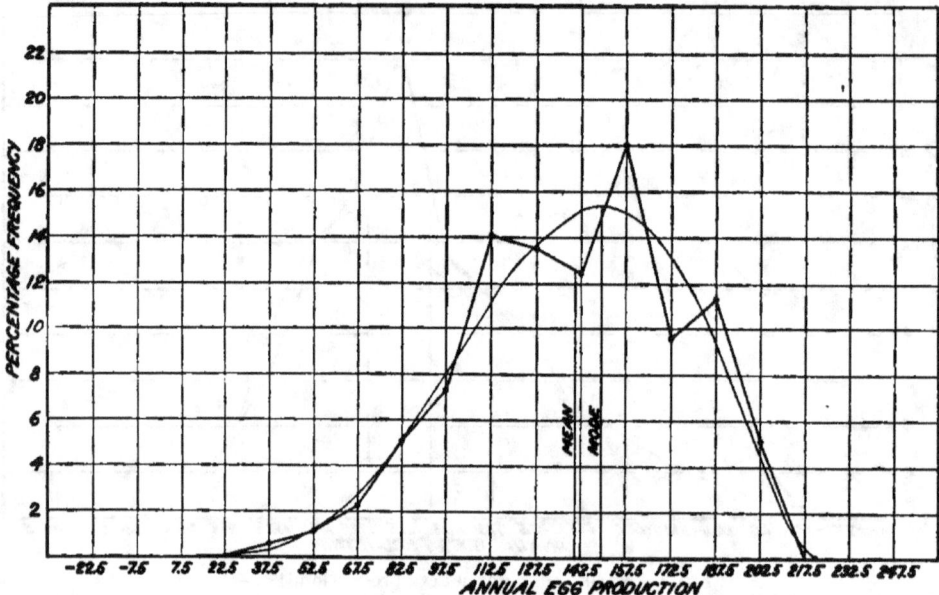

FIG. 3.—Variation in annual egg production, 1905-6, 50-bird pens.

that no hen is capable of laying fewer than no eggs at all in a year, yet, with the single exception noted, the lower range end has a negative value in every case.

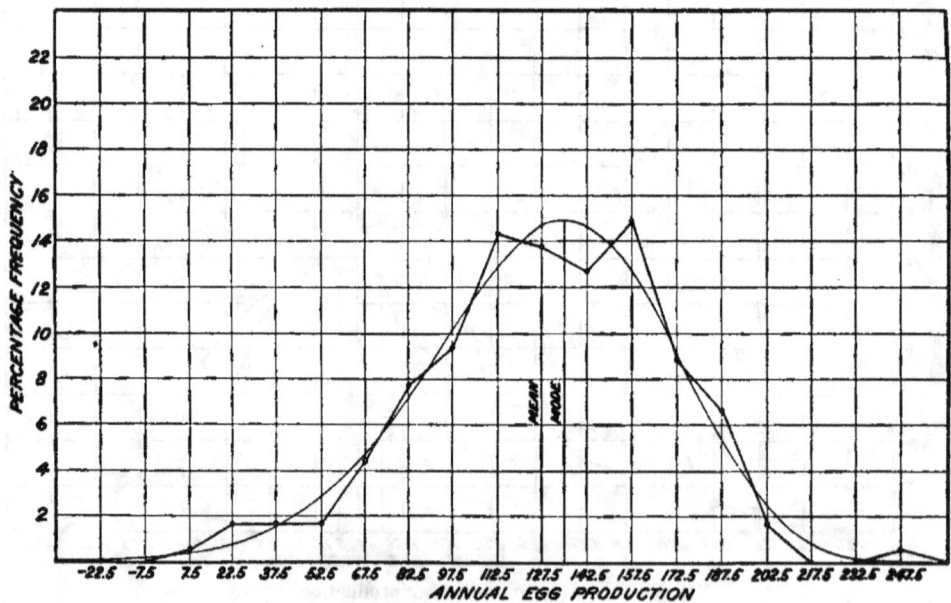

FIG. 4.—Variation in annual egg production, 1905–6, 100-bird pens.

The nearly uniform failure of the fitted curves to give reasonable estimations of the range limits of variation in egg production is remarkable.   Anyone who has had any large experience with Pearson's limited range curves knows that they almost never lead to impossible estimations of the range ends. We have given the matter very careful study in the present case, and while we are able easily to

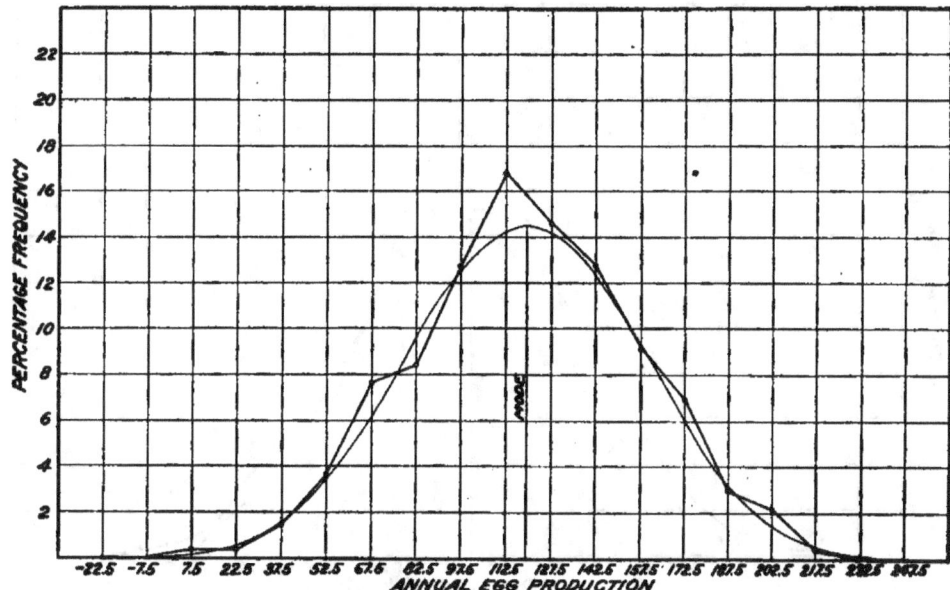

FIG. 5.—Variation in annual egg production, 1905–6, 150-bird pens.

understand and explain the results we get with these curves, we have not been able to modify them in any way so as to get a better representation of the actual range by the theoretical. It occurred to us

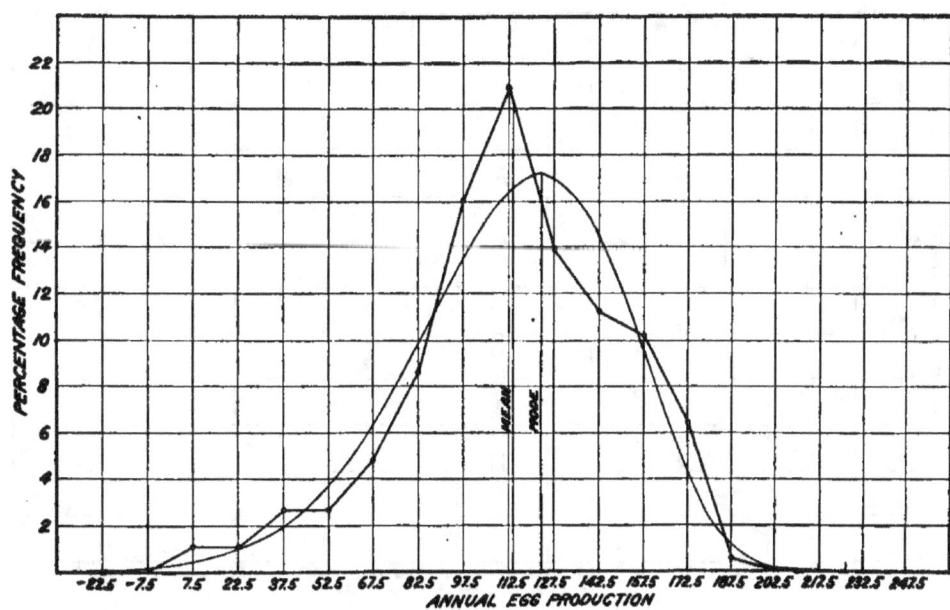

FIG. 6.—Variation in annual egg production, 1906–7, 50-bird pens.

that possibly by using the unmodified moments in the place of those corrected by Sheppard's formulæ we might get a better estimation of the range. Trial, however, proved this not to be the case. In these particular curves it made no substantial difference in the range if Sheppard's corrections were or were not used. It is obvious from the diagrams that the general "fit" of all these curves is excellent.

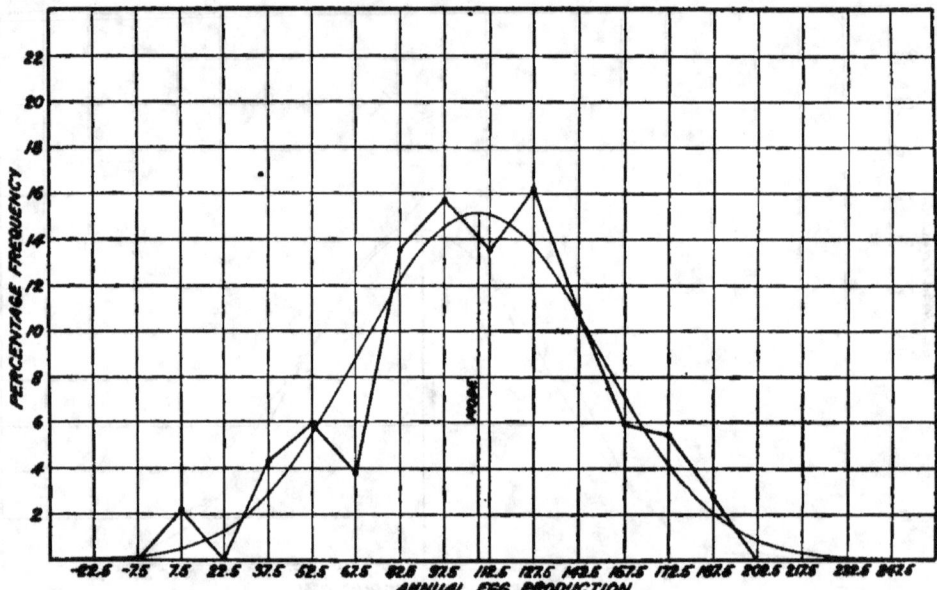

Fig. 7.—Variation in annual egg production, 1906-7, 100-bird pens.

In fact, it would be difficult to get any better graduations than those given by the present curves. Curves of Type I and Type II are the only limited range curves which can possibly be used on these fre-

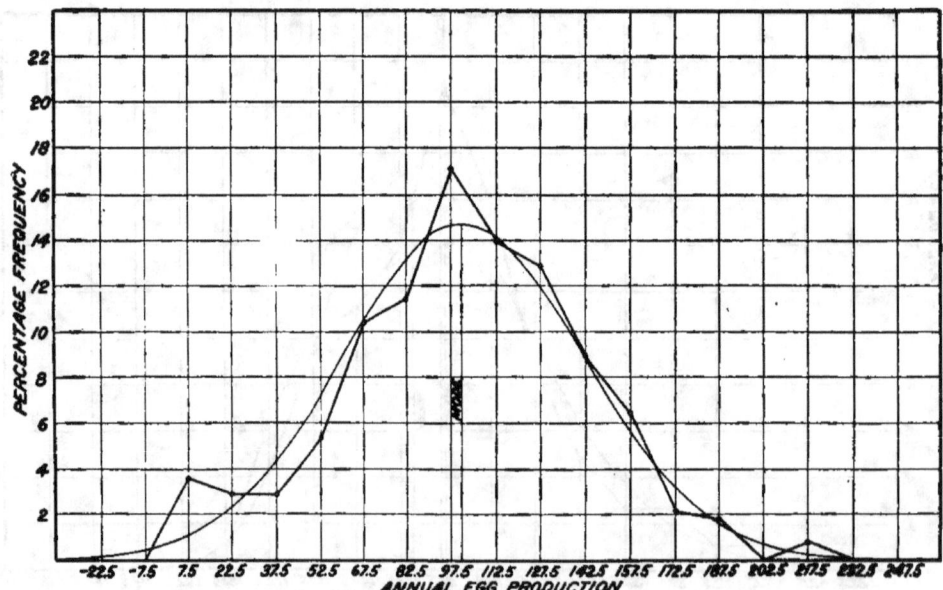

Fig. 8.—Variation in annual egg production, 1906-7, 150-bird pens.

quency distributions, and it is obvious from Table 4 that one of these types is not superior to the other in the matter of the estimation of the range. Hence we can not get over our difficulty by using some other types of curve than those which actually have been used. A remarkable thing is that in every one of the egg-production distributions the value of the criterion $\kappa_1$ is such as to indicate that the distribution deviates from normality in the direction of a limited range curve (either Type I or II). In so far mathematical theory and biological fact agree excellently, because annual egg production is a thing which can only vary within certain limits. The remarkable thing is that while the data demand a limited range curve they lead to impossible values of the range.

Consideration of the matter leads to the conclusion that the chief reasons for the great overestimation of the range by these curves are: (a) That in the majority of cases the curves approach very close to the normal type. The range of the normal curve is from $-\infty$ to $+\infty$. As the limited range curve approaches the normal its range tends very rapidly to increase. (b) That the distributions are in every case rather irregular, owing to the relatively small number of individuals included in each. It will be observed that in each of the worst cases of extension of the range into the negative there is a group of hens laying very few eggs in the year and producing a pronounced hump in the curve near its lower end. Now the theoretical curve attempts to make allowance for this irregularity, but only does so at the expense of sending a long tail off into the negative. With a larger number of observations there is no doubt that much of the irregularity would disappear from the observational polygons, and with it would go that which in considerable measure causes the great overestimation of the range by the present curves.

The points just mentioned should be given due weight before criticism is made of the theoretical curves used because of their failure to give good estimation of the range. In addition the following facts should not be forgotten in the same connection. First, that the total frequency given by the theoretical curves beyond the observed range is in every case so small as to be practically negligible; and second, that the range is subject to a probable error, both in respect to its extent and its position. In the case of the present distributions both these probable errors must be very large for two reasons—one, that the curves approach closely the normal type; the other, that the number of individuals is statistically small.

There is some reason to believe that characters like fertility and fecundity as a rule lead to theoretical curves with considerable overestimation of the range on the negative side. This is certainly the case with our present data. It is true to an even more marked degree in Powys's [a] curves of fertility in human stock. Examination of the data and diagrams which this author gives shows that in spite of

---

[a] Loc. cit.

the large numbers of individuals included in his statistics and the consequent smoothness of the polygons the theoretical curves cross the zero line (no children in the family) with a frequency which is sensible in every case, and which in some cases is relatively much greater than anything we have found in the egg curves.

The general conclusion we reach is that the failure of the theoretical curves to give good estimations of the observed range is to be regarded as the result of peculiarities of the observational data rather than an inherent fault of the curves.

(4) It is of interest to note that in these curves the height of the modal ordinate $y_0$ when expressed as a percentage of the total frequency, is very nearly constant in all cases. Davenport [a] has recently called attention to the significance of the modal frequency as a variation constant.

### CHANGES IN EGG PRODUCTION BETWEEN 1899 AND 1907.

In attempting to reach a definite conclusion regarding the obviously important matter of what has been the character and amount of change in the annual egg production during the period covered by our statistics, we are met at once by the difficulty noted earlier in this paper (p. 19). The data are not strictly comparable from year to year. Improvements were made in feeding, housing, and handling the birds during the period covered by the investigation. Proper allowance can not be made for the influence of these factors. Further, various accidents lowered the egg production in particular years. The only feasible way of getting some definite idea of how the variation constants change appears to be the following: First, to determine the changes in the actual values of the constants as they stand. These may be regarded as lower limiting values. The general trend of these having been determined, we may next proceed to modify the constants, for the influence of accidents and the like, from the actual values they show. In every case we can be sure of the general direction in which such modification should be applied. As to what its exact amount in each case should be there are, of course, no criteria. But, if we make the modification relatively large, we shall establish what may be regarded as a series of upper limiting values for the constants. We may then fairly conclude that had all conditions remained normal throughout the period covered by the data, the values exhibited by the constants would have fallen somewhere between the limits set by the actual and the modified values. While such a method can not give us precise values, it is capable of showing what the general trend of the data is. We may first examine the changes which took place in the mean or average annual egg production in the period from 1899 to 1907.

---

[a] The Principles of Breeding, p. 422.

## CHANGES IN MEAN EGG PRODUCTION.

In considering this problem the question immediately arises as to how the data for 1904 and subsequent years shall be treated. In these years shall we use for comparison with earlier years an average based on all the birds together (i. e., those kept in 50, 100, and 150 bird pens), or only those birds kept in flocks of one size (either 50, 100, or 150)? A little consideration leads to the conclusion that there can be but one answer to this question. We must use the records of flocks of one size only, otherwise the heterogeneity of the material will vitiate conclusions. This being the case, it is further advisable to use the data from the 50-bird flocks, since in the earlier years of the experiment the birds were kept in small flocks, with which the 50-bird flocks of later years are most nearly comparable. Consequently for the years 1904 to 1907, inclusive, we shall use the 50-bird pen records only.

The data for the changes in mean egg production are given in Table 5. The averages given here in the column headed "Actual average production" are those obtained by dividing the figures in the third column of the table by those in the second column. In other words, they are the averages from the ungrouped data given in Table I of Appendix I. On this account they differ slightly from the grouped means of Table 2.

TABLE 5.—*Changes in average egg production between 1899 and 1907.*

| Year and pen. | Birds completing the year. | Eggs laid. | Actual average production. | Added to actual average. | Modified average. |
|---|---|---|---|---|---|
| 1899–1900 | 70 | 9,545 | 136.36 | 0 | 136.36 |
| 1900–1901 | 85 | 12,192 | 143.44 | 0 | 143.44 |
| 1901–2 | 48 | 7,468 | 155.58 | 0 | 155.58 |
| 1902–3 | 147 | 19,906 | 135.42 | a 23.73 | 159.15 |
| 1903–4 | 254 | 29,947 | 117.90 | b 11.24 | 129.14 |
| 1904–5, 50-bird pens | 283 | 37,943 | 134.07 | 0 | 134.07 |
| 1905–6, 50-bird pens | 178 | 24,827 | 140.14 | b 13.95 | 154.09 |
| 1906–7, 50-bird pens | 187 | 21,175 | 113.24 | b 29.53 | 142.77 |

a Cf. page 40.                    b Cf. page 41.

The first point which will be noted from this table is that the averages given in the fourth column do not agree with those which have been published in the several bulletins of the Maine station, in which reports have been given of the progress of the experiment in breeding for increased egg production. Regarding this discrepancy it need only be said that the previously published averages have several times been in error from one or the other of two causes, viz, (a) faulty methods of handling the statistical material, and (b) arithmetical mistakes. Any interested person may verify for himself the averages given in Table 5, from the figures given in Table I of Appendix I, which comprises all the data which exist at the station respecting annual egg production in Barred Plymouth Rocks.

As has been pointed out earlier in the paper (p. 19), more or less serious accidents happened to lower the egg production in four out of the eight years covered in the investigation. The "actual averages" of Table 5 give the lower limit to the egg production of each year. The hens could not have produced fewer than the recorded number of eggs. In the years 1902–3, 1903–4, 1905–6, 1906–7 they probably would have produced more than the recorded number of eggs, had not the various accidents occurred. To obtain an upper limiting value for the egg production of each of these years we must add an allowance for the accident so great that it can only be regarded as unreasonably liberal. Then we can say that under entirely normal circumstances the average egg production could not have been greater than that given by the modified averages. We have then to consider separately the allowances to be added to each year's average.

In the year 1902–3 (cf. p. 16, supra) the birds are stated to have molted in December and to have laid very few eggs in that and the succeeding month. This molting was attributed to too heavy feeding during the summer and early fall months. As a matter of fact the average egg production of the 147 birds was 4.46 eggs in December, 1902, and 5.92 eggs in January, 1903; 76 birds laid in December and 77 in January—a little more than half the flock. Now, a reference to Table 2 will show that the egg production of the 50-bird pens in the year 1905–6 was the highest obtained in any year after 1901–2. Taking into account the number of birds involved it ranks highest, since in the early years there were relatively few birds. In the year 1905–6 the only point at which the laying can be regarded as abnormal is in April and May (cf. p. 18). The egg production in the other months of that year must be regarded as normal. If, then, we add to the 1902–3 average the mean egg production for the whole of the two months of December, 1905, and January, 1906, we shall have made a most liberal allowance for the accident. The mean December, 1905, (50-bird pens) production was 9.90 eggs; that for January, 1906, was 13.83 eggs. We may then add 23.73 eggs to the 1902–3 actual average, giving a modified average of 159.15 eggs. There can be no question that this figure may be taken as a safe upper limit to this year's production.

In the year 1903–4 the only disturbing factor in the records is that the birds were not moved into the houses until December 6. Hence they lack any November record, and also that for the first five days in December. The average production for the month of December was 5.67 eggs. Nevertheless, in order to establish a sufficiently high upper limit to the year's production, we may add to the actual 1903–4 average not only the normal average production for the whole month of November (1905, 50-bird pens), but also one-half of the normal average production of December (1905, 50-bird pens). This gives altogether 11.24 eggs, making the modified average for 1903–4 129.14 eggs. There is no valid reason for adding, as a modifying factor, more than the egg production of (a) the month during which the birds made no records and, of (b) one-half of the month in which they missed five days.

In the year 1905–6 all the conditions were normal, and the egg production was exceptionally high, save for a single mishap, which is said to have affected the April and May totals (cf. p. 18, supra). There is some reason for thinking that only a small allowance should be made for this mishap. The mean production for April is 16.48 eggs and that for May 11.42. In order to be sure of a sufficiently high upper limiting value, however, we may add 50 per cent of the recorded production of each of these months, making the modified April mean $16.48 + 8.24 = 24.72$, and the modified May mean $11.42 + 5.71 = 17.13$. The total modifying factor to be added to the actual mean of this year is then 13.95 eggs, making the modified mean 154.09.

In 1906–7 the application of lice killer to the roosts is stated to have caused a very serious lowering of the egg production in the months of December, 1906, and January, 1907. The actual average egg production (50-bird pens) was for the month of December 4.97 eggs and for January 5.70 eggs. Inasmuch as there is no record of the exact date when the lice killer was applied, we are compelled to add to the actual egg production of 1906–7 as a modifying factor the normal average production for the whole of the months of December and January (based, as before, on 1905 50-bird pen records). A further addition must be made to allow for the fact that this year's records are only for eleven months. There is no record for October, 1907. We may, as in the other cases, add the mean production for this month of the 1905–6 50-bird pens. This gives us the modifying term 29.53 eggs and as the modified average 142.77 eggs.

With the actual and modified averages in hand we may inquire: What has been the general trend of the mean annual egg production during the period covered by the investigation? The clearest answer to this question may be obtained by plotting the figures in the fourth and sixth columns of Table 5, and then striking through each of the two zigzag lines so obtained the best fitting straight line, as determined by the method of least squares. Such a diagram is given in figure 9.

The equations of the two straight lines are as follows:

Actual averages, $y = 148.48 - 3.10\,x$.

Modified averages, $y = 144.13 + 0.043\,x$.

In these equations $y$ denotes the mean annual egg production and $x$ the year. The origin of $x$ is taken at 1898–99—that is, one abscissal unit to the left of the first observation. The reason for so taking the origin is obvious.

Taking the line of the actual averages, it is seen that during the first three years there is a steady rise in mean egg production. The highest mean production during the whole period of eight years is in 1901–2. From this year on the line drops, reaching a very low point in 1903–4. Rising again, the line reaches a second maximum (not so high, however, as the first) in 1905–6. The next year, 1906–7, shows the lowest average production of any year in the experiment. The general trend of the actual averages is obviously downward. This is clearly shown by the fitted line *ab*.

Turning to the line of the modified averages, we note that, while still showing the same individual year fluctuations, its general trend is almost exactly horizontal. As shown by the equation to the straight line *ac*, however, there is an extremely slight upward tendency in the modified averages. It is so small as to be entirely negligible.

Now, the truth must lie between these two lines *ab* and *ac*. One is the graduation of the lower limiting values for each year's average production, the other the graduation of the upper limiting values of

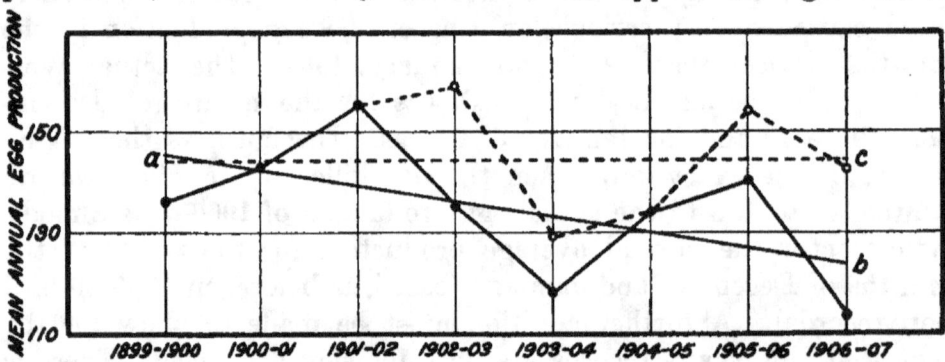

FIG. 9.—Change in mean annual egg production, 1899-1907. The solid lines show the actual averages and best-fitting straight line; the broken lines the modified averages and best-fitting straight line.

what the egg production might have been had not particular accidents occurred. There appears to be but one possible conclusion. In the period from 1899 to 1907 the general trend of average annual egg production has been downward. During the period there have been, as would of course be expected, fluctuations up and down in individual years. These have been no greater, however, than might be expected as purely chance phenomena.

It might possibly be contended that we should reach a different conclusion if we took into consideration all the birds of the years 1904 to 1907, instead of those in 50-bird flocks. A little inspection of Table 2 shows, however, that if the mean productions of all the birds of these last three years were to be used, the general trend of the line representing the change of the mean during the whole period would be more sharply downward than it is in figure 9. This must obviously be the case, because, as the last three lines of Table 2 show, the weighted mean of the means is in each case very considerably lower than the 50-bird pen mean of the same year.

The discussion of the relation of the changes in mean annual egg production to the method of breeding practiced is deferred to a later section of the paper.

### CHANGES IN THE VARIABILITY OF EGG PRODUCTION.

It has been shown that in spite of fluctuations up or down in particular years there is evidence of a gradual but steady change in mean egg production during the period covered by the breeding investigation. Is there any evidence of a similar change in the constants measuring variability in egg production? On a priori grounds we should expect to find a progressive change in variability because of the selection practiced. During the eight generations included in the investigation the immediate ancestry of the birds was strictly selected. On this account we should expect a reduction in variability until a minimum had been reached.[a] Has there been such a reduction?

In attempting to get light upon this matter one is met, as before, by the difficulty that modification ought to be made in the constants for four of the years to allow for the accidents. The matter is not quite so simple in the case of the variation-measuring constants as it is in the case of the means. The problem may be stated as follows: Let $A$ be a character (say egg production during eleven months) having in a given population of individuals a mean value of $m_1$ and a standard deviation of $\sigma_1$, and let $B$ be another character (say one month's egg production) of the same population having a mean given by $m_2$ and a standard deviation given by $\sigma_2$. What will be the mean and standard deviation of the sum of these two characters $A + B$ in the same population?

We have, if $M$ be the mean and $\Sigma$ the standard deviation of the combined characters,

$$M = m_1 + m_2 \quad\text{------------------------------------- (i)}$$

and

$$\Sigma = \sqrt{(\sigma_1{}^2 + \sigma_2{}^2 + 2r_{12}\,\overline{\sigma_1\,\sigma_2}} \quad\text{----------------- (ii)}$$

where $r_{12}$ is the coefficient of correlation between $A$ and $B$.

Equation (ii) presents a number of points of interest and significance. In the first place it is seen that so long at least as $r_{12}$ is positive $\Sigma$ will necessarily be greater than either $\sigma_1$ or $\sigma_2$. Now, as a matter of fact, the egg production of each single month is positively correlated with the total production for the year, as we should obviously expect it to be. That is, $r_{12}$ is in this case always positive. Hence, if we raise the mean egg production of the year by adding as a modifying term the production of one or more months we shall at the same time increase the absolute variability in annual production as measured by the standard deviation.

---

[a] For a full discussion of this matter see Pearson, K., The Law of Ancestral Inheritance. Proceedings of the Royal Society, vol. 62, pp. 386–412.

This leads to another consideration. When the total production of one group of hens is greater in a given time than that of another group in the same time it merely means that the aggregate laying within the period has been increased in one lot over what it is in the other. But, according to (ii) above, this means that at the same time the standard deviation will be greater in one lot than in the other. In other words, when the mean egg production changes we should expect the standard deviation to change in the same direction, unless some real alteration in the innate variational tendency in respect to egg production has occurred. We hence must conclude that if, as in the present case, we find the general trend of mean egg production, and at the same time that of the standard deviation in the same character to be downward, we are not justified in assuming from the standard deviation alone that there has been any progressive change in the innate variational tendency. We may expect the standard deviations to become smaller for the same reasons that the means do. The practical conclusion is that judgment of changes in variation in egg production must be based on a relative measure, such as is given by the coefficient of variation.

Further, if an increase in mean egg production means an increase in the standard deviation except there be a real change in the variational tendency of the population, then clearly it is futile to modify the standard deviations to allow for accidents, as has been done in the case of the means, because such a modification is purely artificial and represents no real change in variational tendency. Hence the coefficient of variation will be unchanged after the modification of mean and standard deviation from what it was before. To show that facts accord with theory on this point an actual case may be cited.

Table 6 gives in parallel columns the values of the chief constants of the distributions for 1904–5, when (a) the records of twelve months' laying (November 1, 1904, to October 31, 1905) are used, and when (b) the totals of the first eleven months of the same laying year (November 1, 1904, to September 30, 1905) are used.

TABLE 6.—*Egg production in 1904–5.*

| Constant.* | 50-bird pens. | | 100-bird pens. | | 150-bird pens. | |
|---|---|---|---|---|---|---|
| | (a) October totals. | (b) September totals. | (a) October totals. | (b) September totals. | (a) October totals. | (b) September totals. |
| Mean | 134.60 | 129.41 | 133.61 | 127.18 | 114.54 | 107.96 |
| Median | 139.17 | 132.90 | 141.00 | 140.25 | 117.00 | 110.67 |
| Mode | 149.14 | 139.71 | 155.78 | 166.38 | 121.93 | 116.08 |
| Standard deviation | 49.49 | 47.25 | 49.41 | 45.90 | 53.04 | 49.56 |
| Coefficient of variation | 36.77 | 36.51 | 36.98 | 36.09 | 46.31 | 45.91 |
| Skewness | −.294 | −.218 | −.449 | −.854 | −.139 | −.164 |

* Probable errors are not given in this table, since they are already given in Table 2 for the constants in the (a) columns, and those in the (b) columns will obviously be of substantially the same magnitude, since the number of individuals involved remains always the same.

From the table it appears that the change in standard deviations made by using twelve instead of eleven month distributions is almost exactly proportional to the change in the means.  The coefficients of variation are not sensibly different in the two cases.  The results then are, first, that it will not be necessary to modify the standard deviations to allow for accidents, and, second, that conclusions respecting variation in egg production must be drawn from the coefficient of variation rather than the standard deviation.

Taking the data given in Table 2, figure 10 has been prepared.  It shows the change in the standard deviations during the eight years covered by the investigation.  In the last three years the constants for the 50-bird pens are used as in the case of the means.  The straight line is the graph of

$$y = 45.12 - 0.88\,x$$

wherein $y$ denotes standard deviation and $x$ number of years from 1898–99.  This is the line which best fits the observations, as determined by the method of least squares.

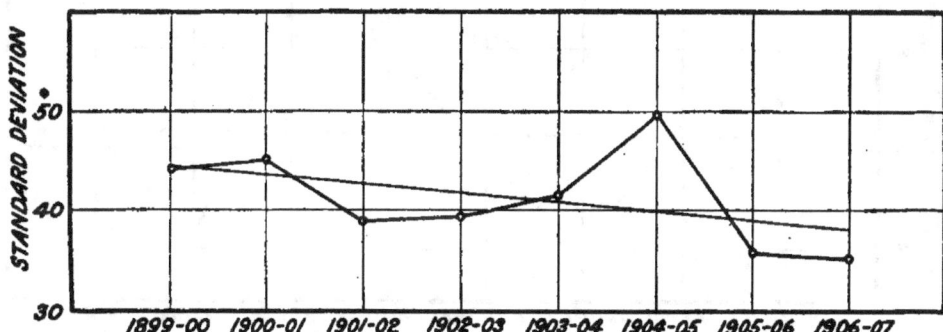

FIG. 10.—Changes in annual egg production.  Standard deviations, 1899–1907.

The general trend of the observations as brought out in this diagram is obviously downward.  However, since the tendency of the means was downward during the same period, it can not be concluded that there has been a real reduction in germinal variability in respect to annual egg production.

Turning to the coefficients of variation and using the data in Table 2 as before, we have the results shown in figure 11.  The constants of 50-bird pens are used in the last three years.

As in the other cases, a straight line was fitted to the observations by the method of least squares.  The equation to the line is in this case

$$y = 30.56 + 0.039\,x$$

where $y$ denotes coefficient of variation and $x$ the number of years from 1898–99.

From this diagram it is seen that the lowest relative variability was shown in the year 1901–2 and the highest in the year 1904–5.

A second minimum, nearly as low as the earlier one, occurs in 1905–6. Finally, the variability in the last year of the period (1906–7) is very nearly the same as that at the outstart eight years before. The fitted straight line is almost exactly horizontal; in fact, the change in the eight years is a little less than 0.3 of a unit. The slight slope which the line does show is upward, but no stress can be laid on so small a change on account of the values of the probable errors concerned. The general conclusion to be drawn is that during the period covered by the statistics the degree of variability in egg production in proportion to the mean has not substantially changed. There have been rather wide fluctuations during the period, but these have been as frequent and great in one direction as in the other.

### CHANGES IN OTHER CONSTANTS.

Nothing in particular is to be gained by a special examination of the changes which have occurred in the median or modal egg produc-

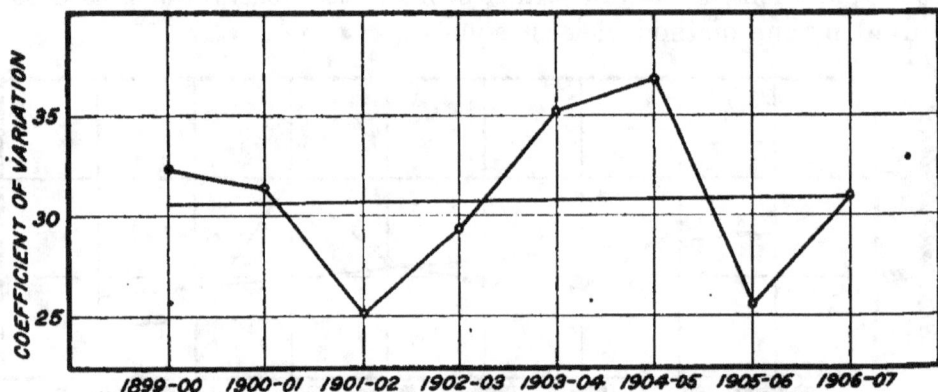

FIG. 11.—Changes in annual egg production. Coefficients of variation, 1899–1907.

tion during the period under investigation. Both these constants run nearly parallel to the means, since the skewness is negative throughout.

It is of some interest to see the trend of the skewness. To this end figure 12 has been prepared. The data are taken from Table 2, and the 50-bird pen constants are used in the last three years. The equation to the straight line is

$$y = -0.265 - 0.0022\ x$$

where $y$ denotes the value of the skewness and $x$ has the same significance as in the other similar equations.

The individual year fluctuations here are rather marked, but there is no evidence of a steady progressive change in the degree or character of the asymmetry of the curves. The fitted straight line is very nearly horizontal.

Another matter of some importance is to determine the proportion of extreme variates in both directions to the total number of birds kept during the several years of the investigation. In this connection the following questions present themselves: In the period from 1899 to 1907 did the number of very exceptionally high-laying hens in proportion to the total number in the flocks increase or diminish, and how great was the change? Similarly, did the porportionate number of exceptionally poor-laying hens increase or diminish, and by what amount? One would expect, of course, in view of the selection practiced in the breeding, that in successive years there would be more "good" layers in the flock and fewer "drones." The data to

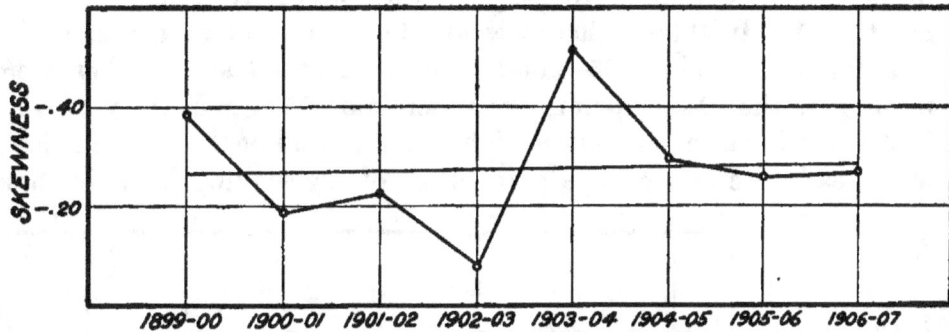

FIG. 12.—Changes in annual egg production. Skewness, 1899–1907.

answer these questions are given in Table 7. We have taken as an exceptionally high-producing hen one that laid 195 or more eggs in her first laying year, and as an exceptionally poor-laying hen one that laid less than 45 eggs in her first laying year. The fairness of these designations can hardly be questioned.

Table 7 gives for each year the percentage (a) of birds laying less than 45 eggs and (b) of birds laying 195 or more eggs. In the last three years data from 50-bird pens are used.

TABLE 7.—*Percentage of the flocks laying (A) less than 45 eggs, and (B) 195 or more eggs in a year.*

| Annual egg production. | 1899–1900. | 1900–1901. | 1901–2. | 1902–3. | 1903–4. | 50-bird pens. | | |
|---|---|---|---|---|---|---|---|---|
| | | | | | | 1904–5. | 1905–6. | 1906–7. |
| | Per cent. | Per cent. | Per cent. | Per cent. | Per cent. | Per cent. | Per cent. | Per cent. |
| Less than 45......... | 4.29 | 1.18 | 0 | 1.36 | 6.70 | 7.07 | 0.56 | 4.81 |
| 195 or more......... | 4.29 | 10.60 | 18.75 | 6.12 | .79 | 12.71 | 5.06 | 0 |

The facts brought out in Table 7 are shown graphically in figure 13, in which the continuous lines show the change in the percentage of exceptionally high layers in the successive years and the broken lines the change in the percentage of very poor layers. The straight lines are the graphs of the following equations, deduced by the method of least squares:

$$\text{Poor layers: } y = 1.795 + 0.3225x$$
$$\text{Good layers: } y = 11.639 - 0.966x$$

In these equations $y$ denotes percentage and $x$ years from 1898–99.

It is noted at once from figure 13 that the changes in the proportionate numbers of exceptionally good layers in successive years very closely parallel the changes in the mean annual egg production as shown in figure 9. In 1899–1900 a little over 4 per cent of the flock laid over 195 eggs in the year. The number increased in 1900–1901 and 1901–2, reaching in this latter year the maximum point of the whole period. This maximum was a high one. Nearly 19 per cent of the flock laid 195 eggs or over. The next two years witnessed a marked decrease, the percentage of high layers reaching in 1903–4 less than 1. In 1904–5 there was another rise in the percentage, with a falling off again in 1905–6 and 1906–7. In the last year there were no birds in the 50-bird pen flocks which laid 195 eggs in the year. As is shown by the continuous straight line, instead of there having been any steady increase in the percentage of high-laying birds in these

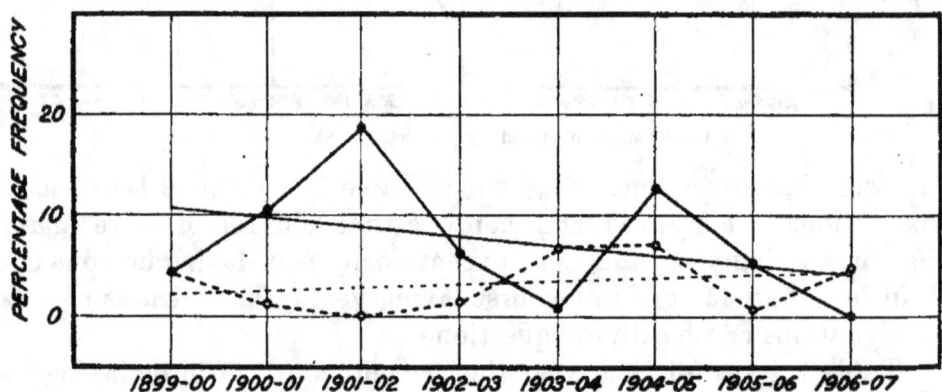

Fig. 13.—Changes in percentages of extreme variates in egg production, 1899–1907. The solid line shows the percentage of birds laying over 195 eggs; the broken line the percentage of birds laying under 45 eggs.

eight years there has, on the contrary, been a distinct decrease in the percentage.

Turning to the poor layers, it is seen that in 1899–1900 there were exactly the same number laying less than 45 eggs as of those laying more than 195. The percentage fell to zero in 1901–2, but during the next three years steadily rose until in 1904–5, 7 per cent of the flock laid fewer than 45 eggs in the year. In 1905–6 there was a drop, with a return in the last year of nearly 5 per cent. Taking all the figures together, it is obvious that the only conclusion which can be drawn is that at the beginning of the selection experiment there were relatively few extreme "drones" in the flock; all the available evidence indicates that after eight years of selection there are as many or more such than there were at the beginning.

Further discussion of the significance of these changes in the number of extreme variates is deferred until a later section of the paper is reached.

## VARIATION IN EGG PRODUCTION IN WHITE WYANDOTTES.

Up to this point we have discussed the variation in annual egg production of the Barred Plymouth Rocks only. As was pointed out in the introduction, the Maine station egg records also furnish some data on the same matter for White Wyandottes. To the discussion of these data we may next turn.

The raw material for the White Wyandottes is given in Appendix I, Table II. The grouped frequency distributions (absolute and percentage) are given in Table 8.

TABLE 8.—*Frequency distributions for variation in annual egg production of White Wyandottes.*

| Eggs laid. | 1899–1900. | | 1900–1901. | | 1901–2. | |
|---|---|---|---|---|---|---|
| | Absolute frequency. | Percentage frequency. | Absolute frequency. | Percentage frequency. | Absolute frequency. | Percentage frequency. |
| 0 to 14 | 1 | 1.43 | | | 1 | 3.03 |
| 15 to 29 | 1 | 1.43 | | | | |
| 30 to 44 | 1 | 1.43 | 1 | 1.39 | | |
| 45 to 59 | | | 2 | 2.78 | 1 | 3.03 |
| 60 to 74 | 5 | 7.14 | 5 | 6.94 | | |
| 75 to 89 | 5 | 7.14 | 6 | 8.33 | 2 | 6.06 |
| 90 to 104 | 6 | 8.57 | 4 | 5.56 | | |
| 105 to 119 | 9 | 12.86 | 7 | 9.72 | 5 | 15.15 |
| 120 to 134 | 11 | 15.71 | 8 | 11.11 | 7 | 21.21 |
| 135 to 149 | 7 | 10.00 | 9 | 12.50 | 8 | 24.24 |
| 150 to 164 | 6 | 8.57 | 12 | 16.67 | 6 | 18.18 |
| 165 to 179 | 10 | 14.29 | 7 | 9.72 | | |
| 180 to 194 | 2 | 2.86 | 4 | 5.56 | 3 | 9.09 |
| 195 to 209 | 4 | 5.71 | 4 | 5.56 | | |
| 210 to 224 | 2 | 2.86 | 1 | 1.39 | | |
| 225 to 239 | | | 2 | 2.78 | | |
| Total | 70 | 100.00 | 72 | 100.01 | 33 | 99.99 |

It will be evident at once from Table 8 and the detailed statistics given in the Appendix that the general features of variation in egg production are very similar in White Wyandottes to what we have already found for Barred Plymouth Rocks. The range of variation is substantially the same in the two cases. As before, the number of observations is in each case too few to fill this range with anything approaching a smooth frequency distribution. It is further apparent that there is a general tendency toward negative skewness in these as in the Barred Plymouth Rock distributions.

The chief physical constants for the distributions of Table 8 are given in Table 9.

TABLE 9.—*Constants of variation in egg production, White Wyandottes.*

| Laying year. | Mean. | Median. | Mode. | Standard deviation. | Coefficient of variation. | Skewness. |
|---|---|---|---|---|---|---|
| 1899–1900 | 130.13±3.64 | 128.50±4.56 | 125.24 | 45.15±2.57 | 34.69±2.20 | +0.108±0.097 |
| 1900–1901 | 135.43±3.54 | 139.50±4.44 | 147.64 | 44.53±2.50 | 32.88±2.04 | −.274±.091 |
| 1901–2 | 130.36±4.37 | 137.50±5.48 | 151.77 | 37.22±3.09 | 28.55±2.56 | −.575±.102 |

From Table 9 we note the following points:

1. The mean annual production of the White Wyandottes is lower in each of the three years than is that of the Barred Plymouth Rocks (Table 2) in the same year. It is extremely doubtful, however, whether this can be taken to represent a true varietal difference from which to conclude that White Wyandottes are poorer layers than Barred Plymouth Rocks. There are several reasons which make such a conclusion too doubtful to carry any weight. In the first place, the number of birds involved is too small. Further, there is no assurance that the same method of feeding is as well suited to egg production in the case of White Wyandottes as in the case of Barred Plymouth Rocks. Further, there is reason to believe that in subsequent years the descendants of the station strain of White Wyandottes came to have a mean annual egg production as high or higher than that of the station Barred Plymouth Rocks. This statement is based on a verbal communication to the writer by Prof. Gilman A. Drew, of the University of Maine, who bred these White Wyandottes after the station dropped them.

2. The White Wyandottes show substantially the same degree of individual variability in annual egg production as do the Barred Plymouth Rocks.

3. In two of the three years the distributions show negative skewness, just as in all cases with the Barred Plymouth Rocks. In the first year the skewness is positive, but its amount is so small and the number of individuals involved is so small that no particular significance is to be attached to this exception to the usual rule for curves of variation in egg production.

The conclusion, so far as any may be drawn from the limited data at hand, is that the general features of variation in annual egg production appear to be substantially the same in White Wyandottes and Barred Plymouth Rocks.

### EGG PRODUCTION IN OTHER BREEDS OF FOWLS.

Reliable and extensive statistical data regarding egg production are exceedingly rare in the literature. Furthermore, whenever any such records have been published it has almost invariably been the custom not to give detailed statistics for individual birds, but rather flock averages based on larger or smaller numbers. From such material it is obviously impossible to learn anything regarding variation in egg production. While this fact is greatly to be regretted, there are certain points of interest to be gained by a comparative study of egg-production averages.

A question which will have occurred to the reader is as to how the average annual productions shown in the different years by the present statistics compare with the experience of others regarding egg production with other breeds of fowls and under different environmental conditions.    Fortunately there is an abundance of data from which light may be gained on this question.    For some years past there have been conducted at various places in Australia egg-laying competitions in which exact records of the number of eggs laid by pens of various breeds have been kept.    The most extensive and thorough of these tests have been carried out by the Hawkesbury Agricultural College and Experimental Farm at Richmond, New South Wales. The data which that institution has accumulated on egg production are of great value.    It is the fashion among poultrymen in this country to discredit these Australian egg-laying competitions and to aver that the published records of their results are wholly untrustworthy. The authors of the present paper have very carefully studied the published reports of these contests and can find no evidence that those conducted at or under the direction of agricultural colleges or government experimental farms (e. g., Hawkesbury and Rockdale competitions) at least, have not been carried out in a strictly scientific spirit and in a manner well calculated both to accumulate accurate scientific data on egg production, and, as judged by the results, to give data which are valuable from the purely practical point of view. We can see no more reason on a priori grounds for repudiating the data from these contests than one would have in repudiating any other quantitative data published by an Australian college or university.

It may be well to state briefly the manner in which these contests were carried out.    The Hawkesbury competitions may be taken as types.    The unit of the competition is a pen of six birds of the same breed.    Intending competitors send the birds they wish to enter at a specified time to the college authorities.    Each pen of six has a separate house and yard.    All yards and houses are of the same size. The birds are cared for throughout the year, and the egg records are kept by an official of the college.    The attempt is made to locate each house and yard in such way that the environmental conditions shall be as nearly as possible identical for all birds in the competition. One would judge from the photographs and descriptions which have been published that this ideal of environmental uniformity has been unusually well attained.    At the inauguration of the first of these contests the following regulations for its conduct were drawn up:[a]

1. The competition to commence on the 1st of April and extend to the 30th of September,[b] a period of six months calendar.

2. The competitors will be bound to pen their birds during March, each pen to consist of six pullets or hens of any age, no male birds to be included.

3. All birds to be bred by the competitor.

[a] Agricultural Gazette of New South Wales, Vol. XVIII, p. 522.  1907.

[b] The reader will not overlook the fact of the reversal of the seasons in the southern hemisphere.  This first contest was a winter egg-laying competition.

4. All birds to be examined by the poultry expert on arrival at the college, and any found to be suffering from infectious or contagious disease to be rejected; in the event of a bird dying, the competitor to be allowed to replace it.

5. All eggs to become the property of the department of agriculture.

6. All competition to be decided by the greatest total number of eggs laid by each pen, eggs under 1½ ounces not to count.

7. The market value of the eggs laid to be recorded and the weight of eggs from each pen, and prizes given for the greatest aggregate weight.

8. Records to be kept of the total quantities of the various feeds consumed and the average cost per head.

After the first contest, rule 1 was changed to include a period of one year in the test. Later rule 2 was also changed in the direction of restricting the age of competing birds, so that the year of the test was the bird's pullet year in each case.

From the standpoint of the scientific study of egg production there is one point open to serious criticism in these Australian statistics. It lies in the fact that no account is kept of the performance of individual birds. The pen instead of the individual is made the unit. This obviously detracts considerably from the scientific availability of the statistics. Another point which is somewhat open to criticism is that involved in the last clause of rule 4, which states that in event of a bird dying the owner may replace it with another. This procedure, however, probably does not seriously affect the averages for comparative purposes. The mortality is stated to have varied in the five years of competition at the Hawkesbury College between 3 and 9 per cent. Further, the plan of replacing each hen that dies with another will give averages for yearly production which we have ascertained to be very closely comparable with those of this paper which exclude the dead, i. e., which count only birds which laid during the whole year. The reason for this is apparent. Suppose a hen dies after having laid during three months. If her record is included with that of birds completing twelve months' laying, its effect will obviously be to lower the flock average. Suppose, however, that when she dies, her place is taken and the twelve months completed by another bird of the same age, and the two are counted as one bird laying twelve months. The effect on the flock average will obviously be to make it very close to what it would have been if the first bird had not died. The only difference which can be produced is that which will arise if the second bird is a distinctively better or poorer layer than the one she replaces. With a small mortality, and any care whatever in the selection of birds, the effect on flock averages must be so small as to be practically negligible. For purposes of comparison with our own statistics, in which only the birds laying the full twelve months are included, it is much better that dead birds should be replaced than not.

The following example will show how small is the effect produced on the mean production by substituting birds for those that die.

From the records for 1905–6 we first chose six birds quite at random. This was done by taking every twentieth bird, beginning with No. 23, i. e., Nos. 23, 43, 63, 83, 103, and 123.   Now, let us suppose that two of these birds, say Nos. 43 and 83, died during the year of the test, and assume that No. 43 died January 31 and No. 83 July 31.   Now assume that the laying year of each of these two birds was completed by another bird taken at random from the flock.   How will the average of the six birds for the year, provided the substitutions are made, differ from what it would have been supposing the two birds, 43 and 83, to have completed their year for themselves?   For the substituted birds two numbers were written down at random.   They were Nos. 494 and 251.   We have the following figures:

| Bird Nos. | Actual laying of birds during whole year. | Until time of assumed death, laying of two birds assumed to have died was— | Substituted birds laid, from assumed time of death to end of year— | Assuming that substituted birds finished year for those dying the laying was— |
|---|---|---|---|---|
| 23 | 152 | | | 152 |
| 43 | 159 | Laying to Jan. 31, 33 | 113 (No. 494) | 146 |
| 63 | 111 | | | 111 |
| 83 | 124 | Laying to July 1, 116 | 27 (No. 251) | 143 |
| 103 | 181 | | | 181 |
| 123 | 147 | | | 147 |
| Total | 874 | | | 880 |

Average production with no death occurring, 145.67 eggs.

Average production assuming that two birds died and that other birds completed the year for them, 146.67 eggs.

The difference between the two averages is obviously insignificant.

For purposes of comparison and illustration some of the data obtained in these Australian contests may be introduced here.

The results of the fifth and most recent Hawkesbury competition[a] which are of significance for our present inquiry are reproduced in Table 10.

TABLE 10.—*Average annual egg production of various breeds.*

[Data obtained in fifth annual egg-laying competition conducted by the Hawkesbury Agricultural College, New South Wales.]

| Breed. | No. of birds. | Average egg production per hen in first laying year. | Breed. | No. of birds. | Average egg production per hen in first laying year. |
|---|---|---|---|---|---|
| Cuckoo Leghorns | 6 | 190.16 | Partridge Wyandottes | 6 | 164.16 |
| Langshans | 18 | 188.88 | Buff Wyandottes | 12 | 163.75 |
| Black Orpingtons | 120 | 178.41 | Buff Leghorns | 18 | 160.55 |
| S. C. Brown Leghorns | 30 | 177.00 | Buff Orpingtons | 18 | 150.11 |
| S. C. White Leghorns | 138 | 174.93 | White Wyandottes | 24 | 146.70 |
| R. C. Brown Leghorns | 12 | 173.50 | Black Leghorns | 6 | 138.33 |
| R. C. White Leghorns | 12 | 172.66 | Houdans | 6 | 137.33 |
| Golden Wyandottes | 12 | 171.33 | Faverolles | 3 | 126.66 |
| Silver Wyandottes | 126 | 170.51 | | | |
| Minorcas | 24 | 168.91 | Total | 597 | 171 |
| Rhode Island Reds | 6 | 166.66 | | | |

a Loc. cit., p. 532.

Similar data from the second Rockdale competition[a] are presented in Table 11, in which the averages given have been calculated from the raw data given in the report of the contest.

TABLE 11.—*Average annual egg production of various breeds.*

[Data obtained in second annual Rockdale egg-laying competition.]

| Breed. | No. of birds. | Total eggs laid. | Average production per hen in first laying year. |
|---|---|---|---|
| Brown Leghorns | 24 | 5,100 | 212.50 |
| White Wyandottes | 18 | 3,765 | 209.17 |
| Andalusians | 6 | 1,255 | 209.17 |
| White Leghorns | 96 | 19,763 | 205.86 |
| Houdans | 6 | 1,196 | 199.33 |
| Minorcas | 6 | 1,174 | 195.67 |
| Black Orpingtons | 96 | 18,459 | 192.28 |
| Silver Wyandottes | 78 | 14,748 | 189.08 |
| R. C. Minorcas | 6 | 1,116 | 186.00 |
| Buff Orpingtons | 18 | 3,097 | 172.06 |
| Langshans | 6 | 774 | 129.00 |
| Total | 360 | [a]70,447 | 195.69 |

[a] This total is incorrectly given as 70,437 in the original table (loc. cit., p. 595).

From these tables it will be noted, first, that an annual egg production very considerably higher than anything observed in the experience of the Maine station is possible even with large numbers of birds. Averages of 205.86 eggs for 96 White Leghorns and 192.28 eggs for the same number of Black Orpingtons (Table 11) are far in excess of any American experience of which the authors are aware. It must, of course, always be kept in mind that these Australian averages are from selected birds, but also (in the case of figures here given) that the selection is made in advance of any direct test of the laying capacity of the selected birds. To be able to select from a flock of pullets which have not yet begun laying a pen of individuals capable of making an annual average such as those we find in these Australian statistics would certainly seem indicative of high laying capacity in the general flock. As a further illustration of this point we may cite a specific instance. In the second Rockdale competition (Table 11) the highest single pen consisted of 6 White Leghorns. In the year they laid 1,473 eggs, making an average of 245.5 eggs per bird. Now, as an examination of Table I of the appendix will show, if we select, a posteriori, the 6 best laying birds which have been found in the whole eight years covered by our statistics, we get the figures on the opposite page.

_____

[a] Loc. cit., pp. 591–691.

|  | Eggs laid. |
|---|---|
| 1899–1900 | 237 |
| 1901–2 | 240 |
| 1904–5 | 234 |
| 1904–5 | 234 |
| 1904–5 | 248 |
| 1905–6 | 246 |
| Total | 1,439 |

These figures give an average per bird of 239.83. Or, in other words, taking the 6 best laying birds that have ever appeared in the Maine station's Barred Plymouth Rock flocks, their average annual laying is nearly six eggs lower than that of the best pen in the particular Australian competition cited. In general we must conclude that the strain of Barred Plymouth Rocks with which we are dealing can not be regarded as an exceptionally high egg-producing one in comparison with other strains and breeds.

A question which constantly recurs to the mind of one studying the problem of egg production is: How does the egg production of the "improved" strains and breeds of the present day compare with that of the domesticated fowl of earlier times?

It is the common opinion that, with the greater attention which has been paid in recent years to poultry raising as a definite and independent branch of agriculture, there has been a great improvement in the so-called "utility" points of the domestic fowl. There have unquestionably been great improvements made in the last twenty-five years in the management of poultry. Equally there can be no doubt that with this improvement in methods of management there has been a decided betterment in the general average condition, in respect to "utility" points, of the domestic fowl. It is not so clear, however, whether within modern times there has been any marked amelioration in the innate qualities on which high egg production depends. This may at first sight appear to be an unwarranted statement. A study of the available evidence, however, we believe, can only lead to the position of doubt on the question just expressed

It should be understood clearly that the question is not as to whether, for example, the average egg production, within a given period, of hens in general is at the present time greater than it was fifty or a hundred years ago. It undoubtedly is. Rather, the significant point is whether, if a given lot of hens of, say a century ago, had been fed, housed, and handled in the same way that our so-called best laying strains are to-day, their egg production would not have been substantially the same as that of the present-day flock. The answer to this question, so far as egg production is concerned, is by no means certain.[a]

The difficulty, of course, lies in getting adequate and trustworthy evidence as to the egg-producing ability of fowls in any but the most recent times. In the older literature one finds plenty of statements that such and such a breed is "noted for its excellence in laying," and the like, but any precise statement as to the criterion of "excellence in laying" then in vogue is almost invariably wanting. Further, when numerical data are given they demand the most careful scrutiny before they can be accepted, because too often it appears from the context that the recorder has some personal interest in making the egg-production record a high one. In such cases the data must obviously be taken with a grain of salt. We have been able, however, to find a few definite records of egg production in earlier times, which appear to be in every respect trustworthy. There is no more reason to doubt their accuracy than there is to doubt the accuracy of any quantitative data of the same period. A few such records may be cited.

In the Journal of Agriculture (English), Volume XI, pages 339 and 340, the following detailed records of egg production are given, on the authority of the editor, in an article on artificial hatching:

*Produce of three Poland pullets hatched in the preceding June, from December 1, 1835, to December 1, 1836.*

| | Eggs laid. | | Eggs laid. |
|---|---|---|---|
| December | 12 | July | 55 |
| January | 50 | August | 55 |
| February | 48 | September | 55 |
| March | 50 | October | 25 |
| April | 54 | November | 9 |
| May | 55 | | |
| June | 56 | Total | 524 |

These figures give as an average production for the year 174.67 eggs per bird, certainly a very creditable performance. We can discover no reason for doubting the accuracy of these figures. They are not presented as in any way exceptional, but rather so far as can be judged from the context, as a definite numerical statement, by way of illustration, of what any poultry raiser might reasonably expect the performance of his hens to be under ordinarily favorable conditions.

---

[a] It is of course certain if the inquiry concerns itself with the so-called "show" points of poultry, such as "shape," "feather," color, and the like. In respect to these characters fanciers have wonderfully improved all the standard breeds. A glance at the plates in such works as Tegetmeier's Poultry Book, for example, affords sufficient evidence of the improvement here.

Another statement of the same general tenor is to be found in an "Essay on the rearing and management of poultry" by William Trotter, published in the Journal of the Royal Agricultural Society in 1851 (vol. 12, pp. 161–202). The essay was awarded a prize by the Royal Agricultural Society in the year mentioned, and hence, we may feel reasonably sure, was subjected to careful and critical scrutiny before publication. It seems, under the circumstances, altogether unlikely that any serious misstatement of fact would have occurred in the essay in the first instance, or if it had occurred, would have been allowed to pass into print. On page 169 (loc. cit.) the following statement is to be found: "Hens of the best laying varieties will lay in a season from 160 to 270 eggs each, averaging 215." If any credence whatever is to be put in this statement, it would appear to indicate that as high egg production as any that we know now was regarded as at least possible fifty years ago.

In another essay which was awarded a prize by the Royal Agricultural Society, and was published in the journal of that society in the year 1867 (series 2, vol. 3, pp. 520–532) a still more definite and precise statement respecting egg production is to be found. The essay is by Mrs. F. Somerville, and its title is "On the rearing and management of poultry on an ordinary farm." The same conclusions as were reached concerning Mr. Trotter's essay appear to apply regarding the trustworthiness of statements in this latter one. On page 532 (loc. cit.) it is stated that in one year "104 hens produced 13,739 eggs *exclusive of those set.*[a] This leads to an average production per hen of 132.11 eggs in the year. The author goes on to state that the 104 hens "reared 372 chickens, besides hatching the ducks and guinea fowls." Now, assuming that it took but 400 eggs to produce 372 chickens, we have the total egg production 14,139 in the year. This gives an average of just under 136 eggs. This can only be regarded as comparing very favorably indeed with the best of our modern records of egg production in equally large flocks, especially when it is considered that these hens had to do all the incubating of at least 372 hen eggs, 79 duck eggs, and 42 guinea-fowl eggs, these being the numbers of each hatched.

We have a number of other references to statements similar to those quoted, but what have been given will suffice for our present purpose. They show that there is trustworthy evidence that at about the middle and the first half of the last century there existed individual birds and flocks of poultry whose ability as egg producers was at least the equal of that to be found in poultry at the present time.

---

[a] Our italics.

## THE INFLUENCE OF CERTAIN HOUSING CONDITIONS ON ANNUAL EGG PRODUCTION.

As has been noted earlier in this paper (p. 17), there has been conducted during the last three years covered by our statistics an experiment to determine the influence of size of flock and amount of floor space per bird on egg production. The general plan of this experiment has already been described and need be only briefly reviewed here. The birds in the experiment were handled in flocks of 50, 100, and 150. The pens were so constructed that in the flocks of 50 and 100 birds there was an allotment of 4.8 square feet of floor space, while in the flocks of 150 birds there was an allotment of 3.2 square feet of floor space per bird. The general plan of construction of the houses in which these birds were kept throughout the course of the experiment has been fully described in earlier publications [a] and need not be taken up again here. The data accumulated in this experiment furnish evidence on two different questions. In the first place, we have data on the influence of the number of birds associated together in one pen on the egg production when the floor space per bird is the same. Evidence of this sort one can, of course, get by comparing the egg records for 50-bird pens and 100-bird pens. In addition there are data on the further question of the influence of amount of floor space per se on the egg production when one brings into the comparison the 150-bird pens.

In conducting this experiment on the influence of floor space on egg production great care has been exercised to make constant all conditions other than housing. The birds used in the experiment have all been of substantially similar breeding. They have been in every case birds bred from mothers laying between 160 and 199 eggs in a year and from fathers whose mothers and earlier ancestors in the female line were birds laying 200 or more eggs in their first laying year. Furthermore, in sorting the birds into the various pens in the fall of each year great care has been taken to make a pro rata distribution of the pullets of different qualities in the different pens. Thus, for example, the pullets on the range would be examined and the very best of them selected, then this selected group would be apportioned in the proper proportion to the different pens in the experiment. The 100-bird pens would receive twice as many of this class of birds as did the 50-bird pens and the 150-bird pens three times as many. Then the second choice would be made of the pullets on the range, and these again divided among the houses in the same proportional manner. This procedure assured as even distribution of birds in the several components of the experiment as probably could in any way be obtained.

---

[a] Maine Agricultural Experiment Station Bulletin 100 and Bureau of Animal Industry Bulletin 90.

It is the purpose at this point to discuss the results of this floor-space experiment only in so far as they have to do with the influence of this factor on annual egg production. In later parts of the work we shall discuss the results of the experiment in respect to the laying in different months of the year, and in other ways shall attempt to make a closer analysis of the data.

We may examine first the influence of the factors size of flock and amount of floor space per bird on the mean or average annual egg production. The data on this matter are presented in Table 12. In addition to the actual means of the different pens for each year there is also given in the last column of this table the excess in the number of eggs produced per bird in the 50-bird pens over that in the 150-bird pens. It will be noted that in this table the actual means are given. It is, for obvious reasons, not necessary to make any modification of any of these averages to allow for the effect of accidents, as was done earlier in the paper in discussing the change in average annual production from year to year. Here the figures which we desire to compare are all based on records of the same year; hence there is every reason to suppose that the influence of any accident which may have occurred will be equally distributed in all parts of the data. That is to say, for example, there is no evidence for supposing that either of the accidents which occurred in the years 1905–6 and 1906–7 affected the 50-bird pens in a different way from what they did either the 100 or the 150 bird pens. Consequently we are justified in using the actual means as they stand.

TABLE 12.—*Relation of size of flock and floor space to mean annual egg production.*

| Year. | Mean annual egg production. | | | Excess of 50 over 150 bird pens. |
|---|---|---|---|---|
| | 50-bird pens. | 100-bird pens. | 150-bird pens. | |
| 1904–5 | 134.60±1.98 | 133.61±3.47 | 114.54±3.02 | 20.06±3.61 |
| 1905–6 | 140.31±1.81 | 127.50±2.03 | 119.43±1.54 | 20.88±2.38 |
| 1906–7 | 114.16±1.74 | 108.53±1.92 | 101.08±1.64 | 13.08±2.39 |
| Mean | 129.69 | 123.21 | 111.68 | 18.01 |

The data in Table 12 are shown graphically in figure 14. From the table and the diagram we note at once the following points:

1. It is obvious that the mean annual egg production is very markedly affected by differences in the environmental factors with which we are dealing. In each year there is a decided difference in the average production of the different pens. Further, it is clear that the general trend of the annual averages is downward as the number of birds in the pen increases. In every instance the mean for the 100-bird pens is smaller than that for the 50-bird pens of the same year, and the mean for the 150-bird pens is again smaller in every case than is that for the 100-bird pens.

2. The decrease in the mean annual production as we pass from the 50-bird pens to the 100-bird pens is, in round numbers, taking the average figures for the whole three years, about 6 eggs. The difference between these two types of pens is not altogether regular from year to year. In the year 1904–5 it is so small as to be insignificant in comparison with its probable error. In 1905–6 the difference is relatively large, amounting to approximately a dozen eggs. In 1906–7 it is about the average, namely, 6 eggs. These results, it will be observed, are perfectly consistent in their general trend throughout the whole course of the experiment, and there appears to be but one conclusion possible. That conclusion is that with the general plan of feeding and housing followed at the Maine station, even

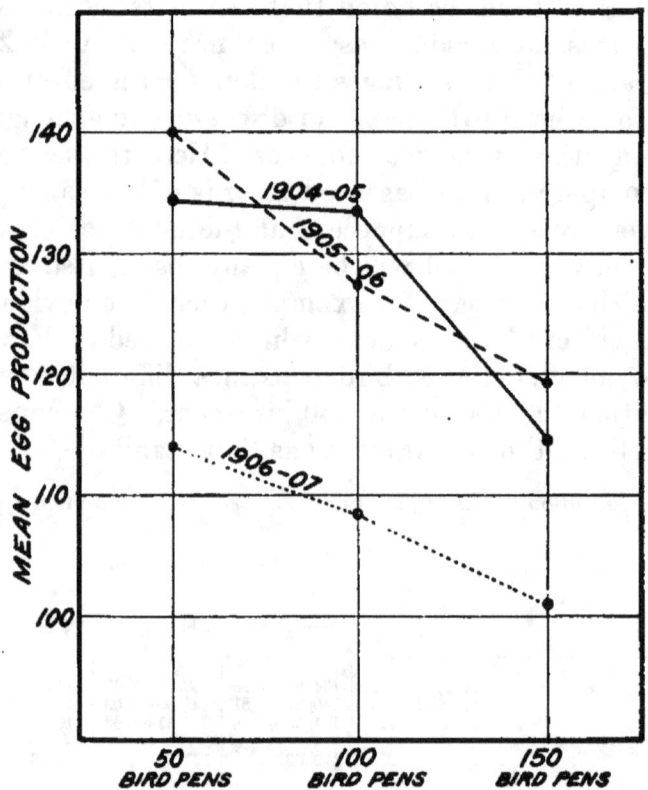

FIG. 14.—Mean annual egg production with different amounts of floor space per bird.

though the amount of floor space per individual bird remains the same, the average egg production per bird in a year's laying is distinctly and significantly smaller when the birds are in flocks of 100 than it is when they are in flocks of 50 birds each.

3. The differences between the averages for 100-bird pens and 150-bird pens are, on the whole, distinctly larger than those between the 50-bird and the 100-bird pens. Taking average figures, the mean annual production is approximately a dozen eggs smaller in the 150-bird pens than it is in the 100-bird pens. As before, the absolute amount of the decrease varies in different years, although the general trend is entirely consistent throughout the course of the experiment.

From these results it must be concluded that under the conditions of the experiment, when we have the combined influence of an increase of number of birds in the flock together with a decrease in the amount of floor space per bird, there is a notable decrease in the average annual egg production.

4. If we consider the maximum difference due to the influence of different amounts of floor space and flock size, as shown between 50 and 150 bird pens, it is seen to be large in amount and relatively constant in different years. In round numbers, the birds in the 50-bird pens averaged to lay a dozen and a half eggs more per bird than did those in the 150-bird pens. Such a large difference is of special interest, as showing what a considerable effect on egg production the environmental factors with which we are dealing may have.

We may next turn to a consideration of the influence of size of flock and floor space per bird on the variability in annual egg production. An examination of Table 2 shows that as the number of birds per pen increases, the standard deviation in egg production also tends to increase. This indicates at once that there is a real increase in the variability of egg production of the large as compared with the small flocks because, as has just been pointed out, the mean egg production decreases as the size of the flock increases. In order to obtain a fair notion of the change in relative variability, we may, as before, resort to the coefficient of variability as the measure. In Table 13 are given the coefficients of variation for each of the different sized flocks and for each of the three years over which the experiment extended. In Table 13, as in Table 12, there is also given in the last column a statement of the maximum differences shown in the pens of different sizes.

TABLE 13.—*Relation of size of flock and floor space to variation in annual egg production.*

| Year. | Coefficient of variation in egg production. | | | Excess of 150 over 50-bird pens. |
|---|---|---|---|---|
| | 50-bird pens. | 100-bird pens. | 150-bird pens. | |
| | Per cent. | Per cent. | Per cent. | Per cent. |
| 1904–5 | 36.77±1.17 | 36.98±2.07 | 46.31±2.23 | 9.54±2.52 |
| 1905–6 | 25.48±0.97 | 31.84±1.23 | 31.72±1.00 | 6.24±1.39 |
| 1906–7 | 30.91±1.18 | 35.54±1.39 | 40.43±1.32 | 9.52±1.77 |
| Mean | 31.05 | 34.45 | 39.49 | 8.43 |

The data of Table 13 are shown graphically in figure 15. From the table and the diagram we note the following points:

1. The relative variability in annual egg production is evidently influenced very decidedly by the environmental conditions. The general trend of the variability, as has been pointed out, is to increase as the size of the flock increases, although there is not so great a degree of regularity in this change as there is in the case of mean egg production.

2. As to the amount of change in variability with the increase in the size of the flock, if we take average figures, it would appear that it is fairly uniform from 50 to 100 and from 100 to 150 bird pens. In each case there is an increase on the average of about 4 per cent in the coefficient of variation. In individual years the increase may not be uniform. For example, in 1904–5 there is practically no difference in the variability in the 50 and 100 bird pens. Again, in 1905–6 the 100-bird pens and the 150-bird pens show substantially the same degree of variation. This apparent equality in variability in egg production in different pens is probably to be explained, however, merely as a result of random sampling, the real tendency being that indicated in the average figures already discussed.

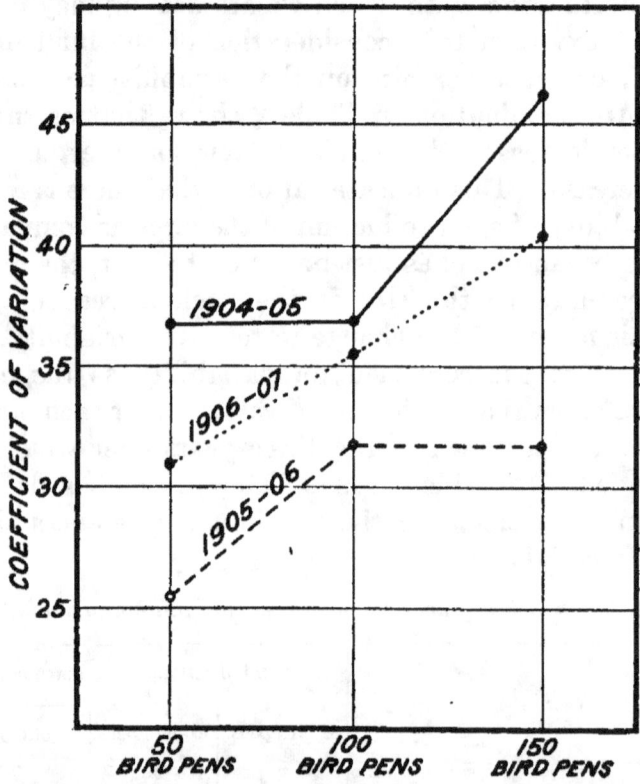

Fig. 15.—Variability in annual egg production with different amounts of floor space per bird.

3. Taking the maximum differences in variability produced by the environmental differences, we see that they are in the neighborhood of 8 to 10 per cent. Since the probable errors of these differences are about 2 per cent, it is obvious that the effect on variation in annual egg production of the environmental factors, with which we are dealing, is an entirely real and definite one. The general result, from present data, would appear to be that an environmental influence which tends to lower the mean annual egg production also tends to increase the variability of the egg production, so that poor-producing flocks are, at the same time, very variable in egg production.

Putting all the results of the floor-space experiment together, they lead to the conclusion that as the size of the flock is increased there tends to be a decrease in the average annual egg production per bird, and this relation holds true whether the floor space per bird is or is not diminished. Further, those conditions relative to size of flock and floor space which tend to cause lower mean annual egg production also tend to increase variation in the production of the flock as a whole. It must be borne in mind that these conclusions apply only to annual egg production, and that from them one must not draw the practical conclusion that it is desirable under all conditions to keep laying hens in small flocks. It will be shown in Part II of this paper, when we come to examine the influence of size of flock and floor space on the egg production during different seasons of the year, that the practical conclusion, relative to the proper floor space and size of flock for laying birds, will be somewhat modified from what one would conclude from the data here presented. For this reason we shall defer drawing any practical conclusions from this experiment or making any suggestions relative to flock management in regard to floor space until the data to be brought out in Part II are at hand.

It will no doubt have occurred to the reader that the results set forth in this section regarding the effect of a strictly environmental factor on egg production have a very significant bearing on the results of the breeding experiment which has been conducted with these birds. It has been shown that a comparatively slight environmental change can in one year produce relatively great changes in mean annual egg production, and, still more important, in the variability of egg production. From this fact alone it would seem doubtful if the method of breeding which has been followed has made any real change whatever in the innate characteristics of the strain relative to egg production. A full discussion of this point will, however, be deferred until a later section of the paper, where the whole question of the influence of selection on egg production will be taken up.

## EGG PRODUCTION IN THE SECOND LAYING YEAR.

Up to this point we have dealt entirely with the egg production in the first laying year, all the egg records ending with the second October following the hatching of the birds. In an earlier section of the paper it has been shown that this first-year egg production may be taken as a fair measure of the innate egg-producing capacity of a bird. It is, however, a matter of some interest to study egg production during the later portion of a hen's life. Unfortunately the data available for such a study are exceedingly meager. In the records of the Maine station there are returns of more than one year's laying for less than a hundred birds. Furthermore, these records of the laying during more than one year are not all that could be desired in another respect. They are records of selected birds and not of a random sample of any particular flock. The birds chosen to be carried through more than one year and subjected to trap-nest records are in most instances birds which laid unusually well during their first laying year.   As a matter of fact, as will appear later, the mean egg production of these birds in their first laying year was higher than any other average production which has ever been observed at this station.   This, of course, is merely the result of selecting especially high-laying birds.   The available data on egg production in the first two laying years have already been published.[a]   From these published data have been taken only the birds which are recorded as having laid some eggs, at least, in both of the first two laying years.   This has given for the present discussion the records

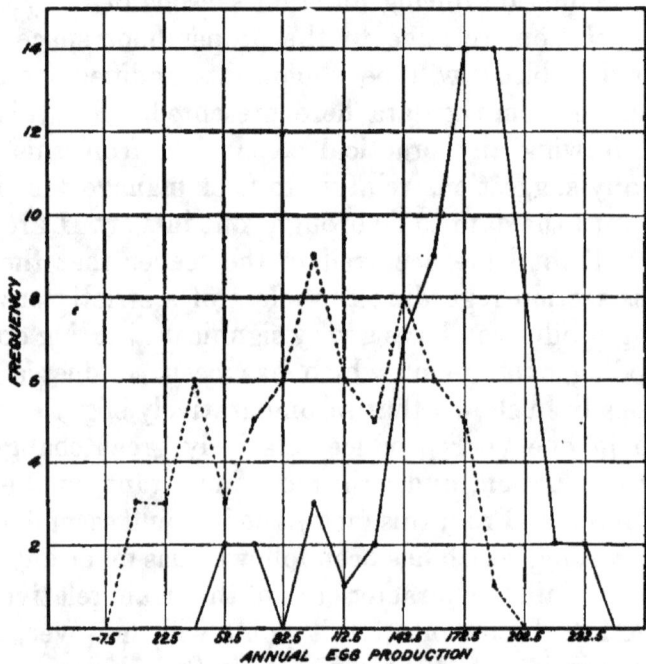

FIG. 16.—Frequency polygons for egg production during the first and second years of laying. The solid line gives the record of the first year, the broken line those of the second year.

of the egg laying for two years of 66 birds.   The egg records for the two years of these birds are given in Table III of Appendix I.

By grouping the egg production as in the other cases into class units of 15 eggs each and forming separate frequency distributions for the first and second years' laying, we have the results shown graphically in figure 16.

The number of individuals involved is so small that the variation polygons are rather irregular.   However, they suffice to show the general trend of the variation in these two periods.   It is noted at once that the variation polygon for the first year's laying possesses the same general characteristics as the others which have been discussed earlier in the paper.   Further, as is of course to be expected,

[a] Maine Agricultural Experiment Station Bulletin 93, p. 75.   1903.

the diagram shows that the second year's production is on the average very decidedly smaller than that of the first year. The material shows in the first year an observational range of variation of 190 eggs—the lowest-yielding bird of the lot having laid 47 eggs during the year and the highest 237 eggs. In the second laying year of the same individuals the lowest-yielding bird laid only 8 eggs during the year and the highest-yielding bird 185 eggs, giving a total observed range of 177 eggs.

While the polygons are much too irregular, on account of the relative fewness of the individuals involved, to enable any final conclusion to be drawn regarding such matters as skewness, yet it seems to be indicated that the polygon of variation for the second year's laying is much more symmetrical about the mean than is that for the first year's laying. The first year polygon is clearly negatively skew, like the great majority of the other egg-production frequency distributions which we have examined.

Turning to the constants of variation in egg production for the two years, we have the results set forth in Table 14.

TABLE 14.—*Constants of variation in first and second year egg production.*

| Year. | Mean. | Standard deviation. | Coefficient of variation. |
|---|---|---|---|
| First laying year | $162.95 \pm 3.30$ | $39.71 \pm 2.33$ | $24.37 \pm 1.50$ |
| Second laying year | $100.47 \pm 3.99$ | $48.06 \pm 2.82$ | $47.84 \pm 3.38$ |

The first point to be noted from Table 14 is the relatively high mean production for the first year—in round numbers, 163 eggs. As has been pointed out, this is due merely to the fact that it is based on a group of birds selected for their egg production and not on a random sample of any year's flock. In the second year, in this particular case, the mean production decreased by rather more than a third of what it was in the first year. Whether the difference between the laying of the first and second years would be the same in a random sample of birds it is impossible now to say. We hope later to get some definite evidence on this point.

It is obvious from the table that, however it is measured, the second year's egg production is very much more variable than is that of the first year. The standard deviation for the second year's production is nearly 10 eggs greater than is that for the first year's. Since at the same time the mean egg production has decreased, it is obvious that the relative variability, as measured by the coefficient of variation, must be very much greater in the second year. This is in fact the case. The relative variability in the second year is nearly twice as great as it is in the first. It will be noted that the variability of this group of 66 birds, in spite of the fact that they were a selected lot, is not widely different from what we have found for samples not especially selected.

A point of much interest is to determine what degree of correlation exists between the laying of the first and the second years. If a bird is an especially good layer in its pullet year will it, on the average, be an especially good layer in its second laying year; or, conversely, if it is a very poor layer in the pullet year will it, in the average case, be an exceptionally poor layer in its second year of life? The only way in which a definite answer can be obtained to this question is to measure the degree of correlation which actually exists between the first and the second year's laying, and then see whether the coefficient of correlation so obtained is significant in comparison with its probable error. Accordingly, we have calculated this coefficient for our present material. On account of the relatively small number of birds represented in the statistics we have not grouped the material in any way in making the calculation. For this material the correlation between the first and the second year's egg production is given by a coefficient,

$$r = 0.032 \pm 0.083$$

while this correlation is positive, yet it is extremely small and obviously insignificant in comparison with its probable error. We can only conclude that so far as any evidence from our present material is concerned there is no sensible correlation between the egg production of the first laying year and that of the second.

We may next consider the characteristics of variation in the combined egg production of the first and second laying years. This is a point of practical as well as theoretical interest, because it has been contended that the proper measure of a bird's producing capacity is the laying of the first two years rather than that of the first year. At the present time there are being conducted in Australia egg-laying competitions in which the egg production of the first two years is taken as the unit. If we add together for each of our 66 birds the production in the first and the second year and then form a frequency distribution, using a class unit of 40 eggs, we get the distribution given in Table 15. In addition to the frequency distribution itself, there are also included in Table 15 the values of the chief physical constants deduced from the distribution.

TABLE 15.—*Distribution and constants of variation in combined first two years' egg production.*

| Eggs laid. | Frequency. | Eggs laid. | Frequency. |
|---|---|---|---|
| 80–119 | 3 | 360–399 | 4 |
| 120–159 | 3 | | |
| 160–199 | 5 | Total | 66 |
| 200–239 | 12 | | |
| 240–279 | 15 | Mean | 263.42±5.84 |
| 280–319 | 12 | Standard deviation | 70.29±4.13 |
| 320–359 | 12 | Coefficient of variation | 26.68±1.67 |

It is obvious that the general characteristics of variation in the egg production of the first two years combined are very similar, so far as may be judged from the present material, to what has been found for the production of the first year alone. The frequency distribution is apparently unimodal and negatively skew. The comparatively low mean egg production given by the two-year distribution is obviously to be explained by the fact just brought out that there is no sensible correlation between the egg production of the two years. Obviously, if the bird which was a relatively high producer in its first laying year was also always or usually a relatively high producer in the second laying year we should get a higher mean egg production in the same lot of birds than that which we have here. With the increase in the mean egg production over that of a single year is found, as would be expected, an increase in the standard deviation. This increase, however, is almost precisely proportional to that which occurs in the mean; hence we are compelled to conclude that the relative variability in respect to the first two years' egg production is not substantially different (as is indicated by the coefficient of variation) from that of the first year alone.

So far as may be judged from such meager data as we now have, there seems no particular reason for regularly taking as a unit in egg-production investigations or contests two years' laying instead of one year's. The important characteristics of variation appear not to be essentially different for the longer unit from what they are for the shorter. Such a conclusion can for the present, however, be only provisional.

It is much to be regretted that more extensive material on this matter of egg production in years subsequent to the pullet year is not at hand. What we have attempted to do in this section is to discuss the very meager material which now exists. It is a part of the plans for future poultry work at the Maine station to make further study of this point, and we hope that later we may be able to present a fuller and more adequate discussion of the matter.

## SELECTION AND EGG PRODUCTION.

We may now turn to the consideration of the results of the breeding experiment which gave origin to the statistics on which this paper is based, so far as these results concern the annual egg production. An account of the experiment has been given earlier in this paper (pp. 11, 12). Its underlying idea was to improve the egg-laying quality of a particular strain of hens by breeding only from the best producers and their sons. It should be said that until the last year of the experiment (1906-7) practically no records were ever kept by which it could be told what birds were the parents of any particular individual offspring. It was simply known that only eggs of high-producing hens were incubated and that the males that fertilized these eggs were sons of high-producing mothers.    In a very few cases records were kept of the mothers of individual birds, and in still fewer cases of the fathers.    These facts will make it plain why it is necessary to discuss the results of this breeding experiment in the manner which will be followed in this section.    The lack of the most fundamental data of all in any breeding investigation—namely, the pedigree data— makes it impossible to attack the problem of the influence of selection on egg production in the usual straightforward fashion.    Obviously, the proper way to study the matter would be to form correlation tables between parent and offspring in regard to egg production and then evaluate the degree of intensity of inheritance of this character, compare the variability in successive generations following selection, etc.    It is impossible to proceed in this way, since we have only mass data instead of individual data.    The only conclusions which can be reached must be such as can be deduced from mass data.

We may first consider the question of the degree of intensity of the selection which has been practiced.    In the several accounts of the breeding experiment [a] it is stated that only birds which laid 160 or more eggs in their pullet year were ever used as breeders.    Now, while at first thought this may seem like an entirely uniform pro- cedure, a little consideration shows that it is far from being such, so far as intensity of selection is concerned.    All selection must be of or about a type.    This is equally true whether the one making the selection has a definite and fixed type in mind and picks out individ- uals as nearly like it as possible, or, as in the present case, the selection is made by truncating the general population distribution and taking as the selected individuals all those falling in the portion cut off. One thinks of selection as most intense when the selected individuals conform closest to type.    The most intense selection possible would be to take out of a population only those individuals which exactly conform to the type within the limit of the unit of measurement. Thus, in the present material, suppose it to be agreed that the type to which selection was to be made was an egg production of 180 eggs in the year.    Then we should get the most intense selection possible by taking as breeders only birds laying exactly 180 eggs.    As birds having productions differing from 180 eggs were taken, we should be selecting with less and less stringency.

These considerations indicate at once a method whereby one may measure with precision the intensity or stringency of the selection applied in any particular case.    From what has been said it is apparent that the smaller the variation exhibited by the selected stock in comparison with that exhibited by the general population from which the selection is made, by so much is the selection more

---

[a] See particularly Bureau of Animal Industry Bulletin 90.

intense or stringent. Hence, if one takes the ratio of the standard deviation of the selected stock to the standard deviation of the general population, the result will be a coefficient measuring intensity of selection. If we use for this coefficient the letter $S$ we have

$$S = \frac{\sigma_M}{\sigma_P}$$

wherein $\sigma_M$ denotes the standard deviation of the selected stock and $\sigma_P$ that of the general population. Now obviously the limits of this coefficient $S$ are 0 and $+1$. When the selection is as stringent as possible and each selected individual is exactly like the type (within the limits of the unit of measurement), there will be no variation in the selected stock, and hence $\sigma_M$ will equal zero, and consequently $S$ will also. If on the other hand the selected individuals are chosen at random, they will approach no more closely to the type than do the individuals of the general population. In that case we shall have $\sigma_M = \sigma_P$ and $S = 1$. Between these two limits, 0 and 1, $S$ will vary in magnitude inversely as the intensity of the selection. We may for convenience call $S$ the coefficient of selection.

In the breeding experiment on which we are here reporting, only hens laying more than 160 eggs in their pullet year were used for breeding. Of such hens, so far as we are able to learn, practically all were used. There were a few rejections because of small or badly colored eggs, but the number of such birds was very small. As has been pointed out above (p. 67), it is now impossible to tell exactly what birds were used as breeders in earlier years. Consequently, we can not give absolute figures for the intensity of selection. What can be done, however, is to determine the intensity of selection in each year of the experiment, assuming that all birds laying 160 or more eggs in their pullet year were used as breeders. The results so obtained, while not absolutely correct, will come very close to the true facts. For all practical purposes one may consider the figures to give the actual intensity of the selection practiced. In Table 16 it is assumed that all birds laying over 160 eggs in the pullet year were mothers of the flock of the second year following their hatching.

TABLE 16.—*Intensity of selection practiced in the breeding, 1899–1907.*

| Year. | General population mean. | Mother's mean. | Difference in means. | General population standard deviation. | Mother's standard deviation. | Difference in standard deviations. | Intensity of selection. | Percentage of mothers to general population. |
|---|---|---|---|---|---|---|---|---|
| 1899–1900 | 136.36 | 179.83 | 43.47 | 44.03 | 20.59 | 23.44 | 0.468 | 34.28 |
| 1900–1901 | 143.44 | 184.83 | 41.39 | 45.02 | 16.06 | 28.96 | .357 | 42.36 |
| 1901–2 | 155.58 | 188.20 | 32.62 | 38.94 | 20.56 | 18.38 | .528 | 47.92 |
| 1902–3 | 136.48 | 182.62 | 46.14 | 39.34 | 17.86 | 21.48 | .454 | 27.88 |
| 1903–4 | 117.87 | 172.70 | 54.83 | 41.43 | 10.86 | 30.57 | .262 | 15.76 |
| 1904–5 a | 129.07 | 186.87 | 57.80 | 51.30 | 18.78 | 32.52 | .366 | 30.10 |
| 1905–6 a | 128.44 | 174.09 | 45.65 | 39.07 | 13.86 | 25.21 | .355 | 20.94 |
| 1906–7 a | 106.99 | 172.33 | 65.34 | 39.09 | 10.84 | 28.25 | .277 | 8.88 |

a For these years, just as in the earlier ones, all the birds in the flock are taken together in determining the constants in this table. In selecting breeding stock no account was made of the size of the pens from which they came. Hence we have no reason to treat 50, 100, and 150 bird pens separately.

From Table 16 we note the following points:

1. The difference between the mean of the selected stock (mothers' mean) and that of the general population varies rather widely in different years. It is, for obvious reasons, least when the general population mean is highest, and greatest when the general population mean is lowest. This difference ranges in value from about 33 eggs in 1901–2 to 65 eggs in 1906–7, and averages 48.41.

2. Similarly there is considerable fluctuation from year to year in the differences between the standard deviation of selected stock and the general population. These differences range between 18 and 32 eggs, with an average of 26.10.

3. The intensity of selection, as measured by the coefficient, is rather great throughout the whole experiment. In the years when the selection is most stringent the coefficient only drops to a value of about one-fourth, while in the year of least intense selection the value is more than one-half. These figures mean that never in the course

FIG. 17.—Change in intensity of selection, 1899–1907.

of the experiment was the selection more than three-fourths as stringent as it was theoretically possible to make it, and it was usually decidedly less close than that.

4. The general trend of selection during the period was to become relatively more stringent; that is, the value of the coefficient of selection diminished during the experiment. This is clearly shown in figure 17. In this figure the zigzag line shows the fluctuation in the value of the coefficient in the different years. The smooth line is the parabola

$$y = 0.4620 - 0.0062x - 0.0020x^2$$

the constants of which were determined by the method of least squares.

The reason for the decrease in the value of the coefficient of selection is clearly to be found in the fact that there was a steady decrease in the egg production of the flocks during the experiment, as has been shown above (p. 42). As the laying becomes poorer there will be fewer and fewer of the flock producing over 160 eggs. Consequently the variation exhibited in the selected stock will steadily become smaller both absolutely and in proportion to that of the general population (coefficient of selection).

5. The last column of the table shows for each year the proportion of the flock which produced over 160 eggs. As we should expect from what has gone before, this proportion diminishes in the later years of the experiment.

Having examined the method and degree of selection practiced, we may next consider the results. According to current theories regarding the inheritance of fluctuating variations, the following are the primary and fundamental results which one would expect to get in such an experiment as this:

1. A gradual and steady tendency for the mean annual egg production to increase with each successive generation. Since all the ancestry for at least eight generations was selected, it would be expected that at the end of this period there would be a marked gain in egg production over what existed at the beginning.

2. A small but not the less distinct and definite decrease in the variability of egg production during the course of the experiment.

3. A considerable degree of fixity of the improvement gained, such that ordinary environmental fluctuations would not in a single generation cause all that had been presumably gained by selection to be lost.

Now, it has been shown in this paper that, so far as annual egg production is concerned, the actual results obtained in this breeding experiment are as follows:

1. The only change which occurred in the mean egg production, aside from individual year fluctuations, was a small but entirely definite decrease during the eight years (pp. 39–42). Further the percentage of exceptionally high producers in the flocks has somewhat decreased during the course of the experiment, and at the same time the proportion of extremely poor producers has slightly increased (p. 48).

2. The variation in annual egg production has remained unchanged throughout the course of the experiment except for minor fluctuations in individual years (p. 46).

3. There is no evidence that the quality of high egg productiveness is any more fixed in the strain at the end of the experiment than it was at the beginning. During the last three years of the experiment it has been definitely shown that a relatively small environmental change is able to produce a very large difference in the mean egg production in flocks of hens of exactly the same selected ancestry (pp. 58–63). Such a result could not occur if the character had been fixed by the selective breeding.

These results we believe to be of great interest and importance, alike from the standpoint of evolution theory and that of practical poultry husbandry. While in a sense they mark a failure to attain that for which the work was undertaken—namely, an improvement of egg laying by this particular method of breeding—yet they possess much educational value. We have here an extensive selection experiment carried out in the traditional time-honored fashion, but with records of a character which make it possible to appreciate the net results with certainty. These results are not in accord with current theoretical ideas, but they are in remarkable accord with the experience of Nilsson at the Svalöf experiment station in breeding cereals, as recently described by De Vries.[a] This is sufficiently shown by the following quotation. Speaking of the earlier experience of the Svalöf station in breeding cereals by the ordinary method of continued selection, De Vries says (loc. cit., pp. 62 and 63):

The deficiencies of the method could not escape observation. All those which were of purely technical nature could be overcome, and on this side of the problem a high degree of perfection was attained. But there were other weak points which were related to the very foundation principles of the method. Among these, the principal one was that the continuous selection of the best specimens in an arbitrary direction did not lead to improvement in all cases. Quite on the contrary, success was rare; so rare, even, that it could almost be looked upon as an exception. The fact itself was not new, since in Germany, also, only in exceptional cases a real improvement had been obtained. But, of course, usually only successful instances are published, and concerning the remainder ordinarily no evidence is at hand.

In Svalöf, however, where numerous experiments of the same kind, but with different varieties of cereals, were conducted side by side, the fact could not escape observation. Soon the idea suggested itself that if success is an exception, the principle involved in the method can hardly be a valuable one, at least, not one on which the breeder may confidently rely.

The fact that in the present case of long-continued selection for a particular kind of animal production we get results so exactly paralleling those obtained in plant breeding when the same method of selection is practiced is certainly an interesting one. A strong argument of those who have endeavored to controvert the views of De Vries regarding the significance of fluctuating and mutational variation in the creation or amelioration of varieties has been the history of domesticated animals. Is the idea that all improvement of breeds on the animal side has been brought about by the continued selection of small "fluctuating" variations as well grounded as it is commonly supposed to be?

It appears to us to be idle to attempt any explanation at present as to why this breeding experiment resulted as it did. Of the objective results we are certain, and we have endeavored to set them forth clearly and distinctly. But for the interpretation of them we need much additional data which we do not as yet possess. Such

---

[a] De Vries, H., Plant Breeding, pp. xiii and 360. Chicago, 1907.

data we hope in course of time to accumulate.   Beginning with the breeding season of 1908 we shall have an exact individual pedigree record of every individual bird in the station's poultry plant.   These records must ultimately throw light on the meaning of our present selection results.

· The practical conclusion to be drawn from the results of this breeding experiment seems to us to be clear.   It is that the improvement of a strain of hens in egg-producing ability by selective breeding is not so simple a matter as it has been supposed to be.   Nothing could be simpler than breeding from high producers to get high producers.   But if this method of breeding totally fails to get high producers—in other words, if the daughters prove not to be like the mothers in egg production—it can not fail to excite wonder as to whether the simplicity of the method is not its chief (possibly its only) recommendation.   Anyone who makes a thorough first-hand study of an extensive selection experiment carried out, as was this one, by the so-called German method without testing the transmitting power of individual organisms, can not fail to be impressed, we believe, with the fact that the improvement of a race by selective breeding is a vastly more complicated matter than it is assumed to be by those who maintain that one need only to breed from the best to insure improvement.   The supposed "facts" of heredity on which the practical stock breeder (working for utility points) operates are in very large part inferences rather than facts.   What is needed more than anything else for the advancement of the stock-breeding industry in all its phases is an accumulation of definite knowledge of the fundamental principles of the hereditary process.   All breeding operations must be based on the laws of inheritance in organisms.   The practical stock breeder is able to work out the applications of these laws for himself.   What he most needs is broader and deeper knowledge of the laws themselves.   This knowledge must come from thoroughgoing purely scientific investigations.

## SUMMARY OF RESULTS AND CONCLUSIONS.

This paper constitutes Part I of a general study of egg production in the domestic fowl.   It deals with variation in annual egg production.   For the purposes of this investigation "annual egg production" is taken to be the total laying in a year beginning with November 1.   For the most part the present paper is concerned with the egg production of the pullet year.   The data on which the work is based are the trap-nest records kept since 1898 at the Maine Agricultural Experiment Station in connection with an experiment in breeding by selection for high egg production.

The chief results of this part of the investigation are summarily stated in the following paragraphs.   Unless something directly to the contrary is said, all the statements in this summary have to do with the egg production in a single variety, namely, the Barred Plymouth Rock.

1. Variation in annual egg production exhibits the following characteristics: (a) The observed range of variation is from zero to about 250 eggs; (b) the distributions are usually unimodal and unsymmetrical. The asymmetry or skewness, when it exists, is always in the negative direction; that is, the modal egg production is always larger than the mean egg production. (c) The amount of variation in egg production is both absolutely and relatively large. The mean value of the coefficient of variation from all our data is about 34 per cent. (d) Variation in egg production, so far as our statistics show, belongs to the type of continuous, so-called "fluctuating" variation.

2. Analytically considered, such of the variation polygons as are skew are found to conform to Pearson's Type I curve. The symmetrical distributions belong either to his Type II or the normal curve of errors. The range of variation is in nearly all cases greatly overestimated by the theoretical curves. It is pointed out that this appears to be a characteristic of fecundity curves frequently.

3. It is shown that during the period covered by the statistics (1899–1907), which covers practically the whole period of the breeding experiment, there has been, apart from fluctuations up and down in individual years, a small but steady decrease in the mean or average annual egg production.

4. During the same period (1899–1907) the variability in annual egg production has not sensibly changed. There have been chance fluctuations up and down in individual years, but there has been no steady trend toward lower or higher variability. The same statement applies to the skewness of the distributions in the period covered by the investigation.

5. The percentage of extremely high layers (producing more than 195 eggs in the pullet year) in the flock decreased during the period from 1899 to 1907. The percentage of exceptionally poor layers (producing less than 45 eggs in the pullet year) in the flock increased during the same period.

6. The general characteristics of variation in annual egg production in White Wyandottes are essentially similar to those described above for variation in Barred Plymouth Rocks.

7. Evidence from the literature is presented tending to show (a) that it is possible to get average annual egg yields higher than any of those which have been observed at the Maine Experiment Station, and (b) that records exist showing that in exceptional cases average annual egg yields were obtained during the middle and earlier half of last century, which were just as high as any we now know.

8. When the laying hens were kept in flocks of 100 birds each the average annual egg production per bird was distinctly and significantly lower than when they were kept in flocks of 50 birds each, though the number of square feet of floor space per bird was the same in the two cases and all other environmental conditions were made as nearly as possible identical.

9. Laying birds kept in flocks of 150 birds each, and with somewhat less floor space per bird than those kept in flocks of 50 and 100 birds each, have in every case an average annual egg production significantly smaller than that of the birds kept in smaller flocks. It is pointed out that great caution must be shown in drawing practical conclusions from these results relative to housing because we are here dealing only with annual egg production.

10. Those conditions of housing and flock size which tend to lower the mean annual egg production are found to tend to increase the variability of the production, so that poor producing flocks are at the same time flocks very variable in production, and vice versa.

11. Egg production in the second laying year is found, on the basis of rather meager data, to average about a third lower and to be distinctly more variable than that in the pullet year. There is no sensible correlation between the production of the first and second year.

12. During the period covered by the statistics all birds used for breeding have been the offspring of mothers laying 160 or more eggs in their pullet year and of fathers which were the sons of high-producing mothers. It is now possible to determine what have been the results of this extensive experiment in selection. It is shown that the intensity or stringency of selection became relatively greater during the progress of the experiment, though the absolute standard of selection remained the same. It is further shown that there is no evidence that the selective breeding practiced has improved the strain in respect to egg production. On the contrary, the data show that (a) the mean egg production has diminished during the experiment, (b) the variability in egg production has remained unchanged, and (c) in the last years of the experiment relatively slight environmental changes caused very marked changes in the flock productiveness. This is obviously inconsistent with the view that any particular type of egg production has in any way been fixed in the strain by the breeding.

# APPENDIX I.

In the following three tables are given the detailed raw data on which the present paper is based:

TABLE I.—*Egg production of Barred Plymouth Rocks.*

| Eggs produced in pullet year. | Number of birds producing in each year the specified number of eggs. | | | | | | | | | | | | | |
|---|---|---|---|---|---|---|---|---|---|---|---|---|---|---|
| | 1899-1900. | 1900-1901. | 1901-2 | 1902-3 | 1903-4 | 1904-5 50-bird pens. | 1904-5 100-bird pens. | 1904-5 150-bird pens. | 1905-6 50-bird pens. | 1905-6 100-bird pens. | 1905-6 150-bird pens. | 1906-7 50-bird pens. | 1906-7 100-bird pens. | 1906-7 150-bird pens. |
| 1. | | | | | 1 | 1 | | 1 | | | | | | 1 |
| 2. | | | | | 2 | 1 | | 1 | | | | | | |
| 3. | | | | | | | | 2 | | | | | 1 | 4 |
| 4. | | | | | | | | 2 | | | | | | 2 |
| 5. | | | | | 2 | | | | | | | | | |
| 6. | | | | | | | | | | | | 1 | 1 | 1 |
| 7. | | | | | 1 | | | 1 | | | | 1 | | |
| 9. | | | | | | | | | | | | | | 2 |
| 10. | | | | | | | | | | | | | 1 | |
| 11. | | | | | 1 | | | | | | 1 | | 1 | |
| 14. | | | | | | 1 | | | | 1 | | | | |
| 15. | | | | | | | | 1 | | 1 | | | | |
| 17. | | | | | | | | 1 | | | 1 | | 1 | |
| 18. | | | | | | | | 1 | | | | | | 1 |
| 19. | | | | | | 1 | | | | | | | | 1 |
| 20. | | | | | | 1 | | 1 | | | | | | 1 |
| 21. | | | | | | | | 1 | | | | | | 1 |
| 22. | | | | | | | | 1 | | 1 | 1 | | | |
| 23. | | 1 | | | 2 | | | 1 | | | | | | |
| 24. | | | | | | | | | | | | | | 1 |
| 25. | | | | 1 | 1 | 1 | | | | | | | 1 | 1 |
| 26. | | | | | | | | | | 1 | | | | 1 |
| 27. | | | | | | | | | | 1 | | | | |
| 28. | | | | | 1 | | | | | | | | | 2 |
| 29. | | | | 1 | 1 | 1 | | | | | | | | 1 |
| 30. | | | | | 2 | 1 | | | | | | 1 | | |
| 31. | | | | | | | | | | 1 | 1 | 1 | 1 | |
| 32. | 1 | | | | | 1 | | | | | | 1 | 1 | |
| 33. | | | | | | | | | | | | | 1 | 1 |
| 34. | | | | | | 1 | | | | | | | | |
| 35. | | | | | | | | | | | | | | 1 |
| 36. | 1 | | | | 1 | 2 | | | | | 1 | | | 1 |
| 37. | | | | | | 1 | | | | | | | | |
| 38. | | | | | | 1 | | | | | 1 | | | |
| 39. | | | | | 1 | 2 | 2 | | | | | 2 | 1 | 1 |
| 40. | | | | | | 2 | 1 | 1 | 1 | 1 | | | 1 | 2 |
| 41. | | | | | 1 | | | | | | | | 1 | |
| 42. | 1 | | | | 1 | 1 | | | | | | | 1 | |
| 43. | | | | | | 1 | | | | | 1 | | 1 | 1 |
| 44. | | | | | | 1 | | 2 | | | 1 | | 2 | 1 |
| 45. | | | | | | 1 | 1 | 2 | | | | 1 | 1 | 1 |
| 46. | | | | | | 1 | | 2 | | | 1 | 1 | 1 | 1 |
| 47. | | 1 | | | 1 | | | | | | 1 | | 1 | |
| 48. | | | | | 1 | 1 | | | | | 1 | | | 2 |
| 49. | | | | | | 2 | | 2 | | | | | 1 | 2 |
| 50. | | | | | 1 | | | | | | 1 | 1 | 1 | 2 |
| 51. | | 1 | | | 1 | 1 | | | 1 | | 1 | 1 | 1 | |
| 52. | | | | 1 | | | | | 1 | 2 | | | 1 | |
| 53. | | | | | | 1 | | | | | 1 | | 1 | |
| 54. | | 1 | | | 1 | 1 | | 1 | | | 1 | | | |
| 55. | 1 | | | | 1 | | | 2 | | | | | 1 | 2 |
| 56. | | 1 | | | 2 | | | | | | 3 | | | 2 |
| 57. | | | | | 2 | | | | | | | | | 1 |
| 58. | | | | | | | | | | | 1 | | 3 | 4 |
| 59. | 1 | | | | | | | | | 1 | 1 | | 1 | |
| 60. | | | | 1 | | 2 | 1 | 1 | | | 1 | 1 | | |
| 61. | 1 | 1 | | | | | | | | | 1 | | | 3 |
| 62. | | | | | 1 | | | | | 1 | 1 | 1 | 1 | 2 |
| 63. | | | | | 1 | | | | 1 | 1 | 1 | | | 1 |

76

TABLE I.—*Egg production of Barred Plymouth Rocks*—Continued.

| Eggs produced in pullet year. | 1899–1900. | 1900–1901. | 1901–2 | 1902–3 | 1903–4 | 1904–5 50-bird pens. | 1904–5 100-bird pens. | 1904–5 150-bird pens. | 1905–6 50-bird pens. | 1905–6 100-bird pens. | 1905–6 150-bird pens. | 1906–7 50-bird pens. | 1906–7 100-bird pens. | 1906–7 150-bird pens. |
|---|---|---|---|---|---|---|---|---|---|---|---|---|---|---|
| 64 |  |  |  |  | 1 | 2 |  | 1 |  | 3 | 3 | 2 | 1 | 2 |
| 65 |  |  |  |  |  | 1 |  |  |  |  |  |  | 1 | 2 |
| 66 |  |  |  |  |  |  |  |  |  | 1 | 2 | 1 |  | 1 |
| 67 |  |  |  |  |  |  |  |  |  |  | 3 |  |  | 1 |
| 68 | 2 |  |  |  | 2 | 1 |  |  |  |  | 1 |  | 1 | 5 |
| 69 |  |  |  | 1 | 1 |  | 1 |  | 1 |  | 1 |  |  | 2 |
| 70 | 1 | 1 | 1 | 2 |  |  | 1 | 1 |  |  | 2 |  | 1 | 2 |
| 71 |  |  |  | 1 |  | 1 | 1 |  |  |  |  | 2 | 1 | 1 |
| 72 |  | 1 |  |  | 2 |  | 1 | 1 | 1 |  | 1 |  | 2 | 1 |
| 73 |  | 1 |  | 1 | 2 | 1 |  | 1 |  |  | 2 | 1 | 1 | 4 |
| 74 |  | 1 |  |  | 1 |  | 1 |  |  |  | 2 | 1 |  | 2 |
| 75 |  |  |  |  |  |  | 2 | 1 | 1 |  | 3 | 1 | 2 | 2 |
| 76 | 1 | 1 |  |  | 1 | 1 | 1 | 2 | 1 |  | 1 | 2 | 2 |
| 77 |  |  |  |  | 1 | 1 |  | 2 |  |  | 3 | 1 | 2 | 1 |
| 78 |  |  |  |  | 1 |  | 1 | 1 | 2 |  |  | 2 | 2 | 1 |
| 79 |  |  | • |  |  |  |  | 1 | 1 |  |  | 2 | 3 |  |
| 80 | 1 |  |  | 1 | 1 | 2 | 2 | 1 | 1 |  |  | 3 | 3 |
| 81 |  |  |  | 1 | 1 | 3 |  | 2 |  | 4 | 1 | 1 | 3 |
| 82 |  |  |  | 2 | 2 | 1 |  | 2 |  | 3 | 3 | 2 | 3 |
| 83 |  |  |  | 1 | 6 | 1 |  | 1 |  | 2 | 3 | 1 | 5 |
| 84 |  |  |  |  | 1 | 1 | 1 |  | 3 | 1 | 1 |  | 2 |
| 85 |  |  |  |  | 1 |  |  |  |  | 1 |  | 1 |
| 86 |  |  |  | 2 | 1 | 1 | 1 | 2 | 1 | 2 | 1 | 4 | 2 |
| 87 |  | 1 |  |  | 1 | 1 |  | 1 | 4 | 2 | 1 | 1 | 4 |
| 88 | 1 |  |  | 1 |  |  |  | 1 | 2 | 1 | 1 | 3 |
| 89 | 1 |  | 1 | 4 | 2 |  |  | 1 | 3 | 1 | 1 | 1 |
| 90 | 1 |  | 1 | 1 | 2 |  |  |  | 3 | 2 |  | 1 |
| 91 |  | 1 | 2 | 1 |  | 1 |  | 1 | 1 | 4 | 3 |
| 92 |  | 1 | 2 |  | 2 | 2 | 2 | 2 | 6 | 3 | 2 |
| 93 |  | 1 |  |  | 1 | 2 | 5 | 1 | 2 | 3 |
| 94 |  | 1 | 2 | 3 | 1 | 1 | 1 | 3 |  | 1 | 4 |
| 95 |  | 1 | 1 |  | 3 | 1 | 2 | 1 | 1 | 1 |
| 96 |  | 1 | 1 | 1 | 1 | 1 | 2 | 3 | 1 | 3 |
| 97 |  | 1 | 2 | 1 | 1 | 1 | 3 | 3 | 3 | 4 |
| 98 | 1 | 5 | 2 | 1 | 2 | 2 | 3 | 4 |
| 99 | 1 | 1 | 1 | 1 | 2 | 1 | 1 | 2 | 3 | 2 | 1 | 2 |
| 100 | 1 | 1 | 2 | 2 | 1 | 1 | 1 | 1 | 2 | 4 |
| 101 | 1 | 2 | 1 | 2 | 2 | 1 | 2 | 3 |
| 102 | 2 | 2 | 1 | 1 | 1 | 2 | 4 | 1 | 3 |
| 103 | 1 | 1 | 1 | 2 | 2 | 2 | 2 | 3 |
| 104 | 1 | 2 | 4 | 3 | 2 | 2 | 2 | 5 | 5 |
| 105 | 1 | 2 | 2 | 1 | 2 | 1 | 1 | 1 | 4 | 4 | 1 |
| 106 | 1 | 1 | 2 | 2 | 1 | 4 | 2 | 4 |
| 107 | 1 | 6 | 5 | 2 | 2 | 3 | 4 | 3 | 1 |
| 108 | 2 | 3 | 3 | 1 | 2 | 2 | 1 | 1 |
| 109 | 3 | 4 | 1 | 1 | 3 | 2 | 4 | 2 | 3 |
| 110 | 2 | 1 | 4 | 1 | 1 | 2 | 1 | 2 | 2 |
| 111 | 2 | 1 | 2 | 3 | 2 | 2 | 1 | 1 | 7 | 1 | 2 | 4 |
| 112 | 3 | 4 | 1 | 1 | 3 | 9 | 4 | 2 | 2 |
| 113 | 1 | 1 | 2 | 1 | 2 | 2 |
| 114 | 2 | 1 | 1 | 3 | 4 | 3 | 2 | 1 |
| 115 | 1 | 2 | 1 | 3 | 3 | 1 | 3 | 3 | 1 | 1 | 6 |
| 116 | 2 | 3 | 4 | 2 | 3 | 2 | 2 | 5 |
| 117 | 1 | 1 | 2 | 2 | 1 | 3 | 2 | 2 | 5 | 2 | 3 |
| 118 | 1 | 1 | 3 | 2 | 1 | 1 | 3 | 2 | 2 | 4 |
| 119 | 1 | 1 | 4 | 1 | 2 | 1 | 6 | 1 | 2 |
| 120 | 1 | 1 | 1 | 2 | 1 | 1 | 1 | 4 | 3 | 2 |
| 121 | 1 | 1 | 2 | 1 | 2 | 2 | 1 | 3 |
| 122 | 1 | 5 | 1 | 1 | 3 | 1 | 2 |
| 123 | 1 | 3 | 2 | 1 | 4 | 3 | 6 | 2 | 4 |
| 124 | 1 | 2 | 1 | 7 | 1 | 3 | 2 | 6 | 3 |
| 125 | 1 | 1 | 5 | 1 | 1 | 2 | 4 | 2 | 4 |
| 126 | 2 | 3 | 3 | 3 | 2 | 3 | 1 | 2 | 4 |
| 127 | 1 | 2 | 2 | 1 | 5 | 1 | 2 | 1 |
| 128 | 1 | 1 | 2 | 2 | 1 | 2 | 2 |
| 129 | 2 | 2 | 1 | 8 | 2 | 1 | 1 | 1 | 3 | 2 | 1 | 2 | 2 |
| 130 | 2 | 1 | 1 | 1 | 2 | 3 | 2 |
| 131 | 1 | 1 | 1 | 6 | 4 | 1 | 2 | 1 | 2 | 4 | 1 |
| 132 | 1 | 2 | 1 | 5 | 3 | 1 | 1 | 6 | 1 | 2 |
| 133 | 1 | 2 | 3 | 2 | 2 | 2 | 1 | 1 | 9 |
| 134 | 2 | 7 | 2 | 2 | 2 | 6 | 2 | 2 |
| 135 | 1 | 1 | 3 | 2 | 2 | 6 | 2 | 2 | 4 | 2 |
| 136 | 1 | 4 | 4 | 2 | 1 | 1 | 3 | 3 | 2 | 3 | 4 |
| 137 | 1 | 4 | 1 | 3 | 5 | 1 | 1 | 3 | 5 | 1 |
| 138 | 1 | 1 | 3 | 3 | 4 | 1 | 1 | 2 | 6 | 1 |
| 139 | 1 | 2 | 1 | 2 | 3 | 1 | 2 | 2 | 3 |

TABLE I.—*Egg production of Barred Plymouth Rocks*—Continued.

| Eggs produced in pullet year. | Number of birds producing in each year the specified number of eggs. | | | | | | | | | | | | | |
|---|---|---|---|---|---|---|---|---|---|---|---|---|---|---|
| | 1899–1900. | 1900–1901. | 1901–2 | 1902–3 | 1903–4 | 1904–5 50-bird pens. | 1904–5 100-bird pens. | 1904–5 150-bird pens. | 1905–6 50-bird pens. | 1905–6 100-bird pens. | 1905–6 150-bird pens. | 1906–7 50-bird pens. | 1906–7 100-bird pens. | 1906–7 150-bird pens. |
| 140 | | | | 4 | | 4 | 1 | | 2 | 1 | 2 | 1 | 1 | 1 |
| 141 | 2 | 1 | | 1 | 4 | 3 | | | | 2 | 2 | 3 | 2 | 1 |
| 142 | | | | 1 | 3 | 2 | | 2 | 1 | | 2 | 2 | 1 | 4 |
| 143 | 1 | | | 1 | 2 | 1 | | 2 | 1 | 1 | 1 | | 2 | 2 |
| 144 | 1 | 1 | | 2 | 3 | 1 | | 1 | 2 | 2 | 1 | 1 | | 2 |
| 145 | 1 | 2 | | 2 | 1 | 1 | | | 2 | 3 | 2 | 3 | 1 | 1 |
| 146 | | 2 | | 1 | 3 | 2 | 2 | 1 | 2 | | 3 | | 1 | |
| 147 | 1 | | | 1 | 1 | 4 | 2 | | 3 | | 2 | 4 | 1 | |
| 148 | | | 1 | 2 | 3 | 4 | 1 | | | | 2 | | 1 | 1 |
| 149 | 1 | 1 | | 1 | 2 | 4 | | | 3 | | 2 | | 1 | 3 |
| 150 | 1 | | | | | 5 | | 3 | 2 | 2 | 1 | 1 | 2 | 3 |
| 151 | | | | 1 | 3 | | | 1 | 1 | 1 | 1 | 2 | | |
| 152 | 1 | | | 1 | 3 | 4 | | | 4 | 3 | 4 | 1 | 1 | |
| 153 | 2 | | 2 | 2 | 4 | 1 | | | 3 | | 4 | 3 | 1 | 1 |
| 154 | | 1 | 3 | 1 | 1 | 2 | 2 | 2 | 2 | 1 | 1 | 1 | 1 | 1 |
| 155 | | 2 | 3 | 2 | | | 1 | 1 | 3 | 5 | 1 | 1 | 1 | |
| 156 | | | | 1 | 1 | 3 | 2 | 1 | 2 | 3 | 2 | 1 | | 2 |
| 157 | | 1 | 1 | 3 | 2 | 3 | | 3 | 2 | 1 | 2 | | | 3 |
| 158 | 1 | | 1 | 1 | | 4 | 1 | | 4 | 1 | 2 | 2 | 1 | 1 |
| 159 | 1 | | 1 | | 1 | 2 | | 1 | 2 | 1 | 1 | 1 | | |
| 160 | | 1 | | 1 | | | 1 | | 2 | 1 | 4 | | | 4 |
| 161 | | | | 1 | 1 | 2 | | 2 | 2 | 2 | 1 | | 1 | 2 |
| 162 | | | 1 | | 5 | 5 | 1 | | 1 | | | | | |
| 163 | 3 | | 1 | 2 | 3 | 1 | 1 | 2 | 1 | 3 | 1 | 1 | 1 | |
| 164 | | | 1 | 1 | 5 | 1 | 1 | 1 | 2 | | 2 | 5 | 2 | 1 |
| 165 | | 1 | | 3 | 1 | 2 | 1 | | 1 | | 2 | 1 | | 2 |
| 166 | 1 | 1 | | 2 | 3 | 6 | | 2 | 2 | 2 | 2 | 1 | | 1 |
| 167 | 1 | 2 | | 2 | 1 | 1 | 1 | 1 | 3 | 1 | 3 | 2 | | 1 |
| 168 | | | 2 | 1 | 1 | 3 | 1 | 1 | | | 1 | | 4 | 1 |
| 169 | 1 | 1 | | 1 | | 1 | 1 | | 3 | 1 | | | | |
| 170 | 2 | 3 | 1 | 1 | | 2 | 1 | 3 | 3 | 1 | | 1 | 1 | 1 |
| 171 | 1 | | | 2 | 2 | | | 1 | 1 | 1 | | 1 | | |
| 172 | | | | 1 | 1 | 3 | 3 | 1 | 1 | 1 | 3 | 1 | | |
| 173 | | | 2 | | | 1 | | | | 2 | 1 | 1 | | |
| 174 | 1 | 1 | 1 | | 1 | 2 | | | 1 | 1 | 2 | 2 | 2 | |
| 175 | 2 | 1 | | 1 | 1 | | | | | | 1 | 1 | 1 | |
| 176 | 1 | | | 1 | 3 | 5 | | 1 | 1 | 3 | 1 | | | |
| 177 | | 2 | | 1 | 1 | | | 1 | 1 | 1 | 3 | | 1 | |
| 178 | 1 | 1 | | 1 | 1 | 1 | 1 | | | | | 1 | | |
| 179 | 1 | 1 | 1 | 1 | | 1 | 1 | | | 2 | | | 1 | |
| 180 | 1 | | | 2 | 1 | | 1 | 1 | 2 | 1 | 1 | | | 2 |
| 181 | 2 | 2 | | | | 1 | | 1 | 2 | 2 | | | | 2 |
| 182 | | 1 | | 1 | 2 | 1 | 1 | | | | 2 | | | |
| 183 | 1 | 1 | | | 1 | 1 | | 2 | | | 2 | | 1 | |
| 184 | | | | | 1 | 2 | 1 | | 1 | | 1 | | 1 | |
| 185 | 1 | 2 | | 3 | | 3 | 1 | | 3 | 1 | | | | |
| 186 | | | | | 1 | 1 | 1 | 1 | 2 | 1 | | 1 | 1 | |
| 187 | | | | | 1 | | 1 | | 1 | | | | | |
| 188 | | | 1 | | | | | | 3 | 2 | | | | |
| 189 | | 1 | | 1 | 1 | 1 | | | 2 | | 1 | | | |
| 190 | | 3 | | | 1 | 3 | | | 3 | 1 | | | 1 | 1 |
| 191 | | 1 | | 1 | | | | | | 1 | | | | |
| 192 | | 1 | | | | 1 | 1 | | 1 | | 1 | | 1 | |
| 193 | 1 | | 1 | | | 2 | | | | 1 | | | | |
| 194 | | | | | | | 1 | | | 2 | | | | |
| 195 | | 1 | | | | | 1 | 2 | 2 | 1 | 1 | | | |
| 196 | | 1 | 1 | | 1 | | | | 1 | 2 | 2 | | | |
| 197 | | | | | | 1 | 1 | | 2 | 1 | | | | |
| 198 | | | 1 | | | | | | | | 1 | | | |
| 199 | | 1 | | | | | 1 | 1 | 1 | | | | | |
| 200 | | 2 | | | 1 | 4 | | | | | | | | |
| 201 | | | 1 | | | 5 | 2 | 1 | | 1 | | | | |
| 202 | | | | | | 2 | 1 | | | | 1 | | | |
| 203 | | | | | | 1 | 3 | 1 | 2 | | | | | |
| 204 | | | | | | 2 | | 2 | | | | | | |
| 205 | | | 1 | 1 | | 4 | | | | | | | | |
| 206 | | | 1 | | | 3 | | | | | | | | |
| 207 | | | | | | 1 | | | | | | | | |
| 208 | 1 | | | | | 2 | | | 1 | | 1 | | | |
| 209 | | 2 | | | 1 | | 1 | | | | | | | |
| 210 | | | | 1 | | 2 | | | | | | | | 1 |
| 211 | 1 | | 1 | | | 2 | | | | | | | | 1 |
| 212 | | | | | | 1 | 1 | 1 | | | | | | |
| 213 | | 1 | 1 | | | | | | | | | | | |
| 214 | | | | | | | | | | | 1 | | | |

TABLE I.—*Egg production of Barred Plymouth Rocks*—Continued.

| Eggs produced in pullet year. | Number of birds producing in each year the specified number of eggs. | | | | | | | | | | | | | |
|---|---|---|---|---|---|---|---|---|---|---|---|---|---|---|
| | 1899-1900. | 1900-1901. | 1901-2 | 1902-3 | 1903-4 | 1904-5 50-bird pens. | 1904-5 100-bird pens. | 1904-5 150-bird pens. | 1905-6 50-bird pens. | 1905-6 100-bird pens. | 1905-6 150-bird pens. | 1906-7 50-bird pens. | 1906-7 100-bird pens. | 1906-7 150-bird pens. |
| 215 | | | | | 1 | | | | | | | | | |
| 217 | | | | 1 | | 1 | | | | | | | | |
| 220 | | | | 2 | | | | 1 | | | | | | |
| 221 | | | | 1 | | 1 | | | | | | | | |
| 222 | | | 1 | 1 | | | | | | | | | | |
| 225 | | | | 1 | | | | | | | | | | |
| 226 | | | | | | 1 | | | | | | | | |
| 234 | | 1 | | | | 2 | | | | | | | | |
| 237 | 1 | | | | | | | | | | | | | |
| 240 | | | 1 | | | | | | | | | | | |
| 246 | | | | | | | | | | 1 | | | | |
| 248 | | | | | | 1 | | | | | | | | |
| Total... | 70 | 85 | 48 | 147 | 254 | 283 | 92 | 140 | 178 | 182 | 275 | 187 | 185 | 281 |

TABLE II.—*Egg production of White Wyandottes.*

| Eggs produced in pullet year. | Number of birds producing in each year the specified number of eggs. | | |
|---|---|---|---|
| | 1899-1900. | 1900-1 | 1901-2 |
| 1 | | | 1 |
| 10 | 1 | | |
| 22 | 1 | | |
| 36 | | 1 | |
| 41 | 1 | | |
| 52 | | 1 | |
| 54 | | 1 | |
| 55 | | | 1 |
| 60 | | 1 | |
| 61 | 1 | | |
| 62 | 1 | 1 | |
| 64 | | 1 | |
| 65 | | 1 | |
| 66 | 1 | | |
| 68 | 2 | | |
| 72 | | 1 | |
| 78 | 2 | 1 | 1 |
| 80 | | 1 | |
| 81 | | 1 | |
| 82 | | | 1 |
| 83 | 1 | | |
| 86 | | 1 | |
| 87 | 2 | 2 | |
| 90 | | 2 | |
| 93 | 1 | | |
| 98 | 1 | | |
| 99 | 1 | 1 | |
| 100 | 1 | | |
| 103 | | 1 | |
| 104 | 2 | | |
| 107 | 2 | 1 | 1 |
| 108 | 2 | 1 | 1 |
| 109 | | | 1 |
| 110 | | | 1 |
| 113 | 1 | 1 | |

| Eggs produced in pullet year. | Number of birds producing in each year the specified number of eggs. | | |
|---|---|---|---|
| | 1899-1900 | 1900-1 | 1901-2 |
| 115 | 1 | | |
| 117 | 1 | | |
| 118 | 1 | 1 | |
| 119 | 1 | 3 | 1 |
| 120 | 2 | | 2 |
| 121 | | 1 | |
| 124 | | 2 | |
| 125 | 1 | | |
| 126 | 2 | | |
| 127 | 1 | | |
| 128 | 2 | 2 | 3 |
| 129 | 1 | | 1 |
| 130 | 1 | | 1 |
| 131 | | 1 | |
| 132 | 1 | 1 | |
| 134 | | 1 | |
| 136 | 1 | 1 | |
| 137 | | | 1 |
| 138 | | 1 | |
| 139 | | 1 | 1 |
| 140 | | 1 | 2 |
| 141 | 1 | 1 | |
| 142 | 2 | 1 | |
| 143 | | | 2 |
| 144 | 1 | 2 | |
| 146 | | 1 | |
| 147 | 1 | | |
| 149 | 1 | | 2 |
| 150 | | 1 | |
| 152 | | | 1 |
| 153 | 1 | 1 | |
| 154 | | 1 | |
| 155 | 1 | | |
| 156 | 2 | 1 | 1 |
| 159 | | 3 | |

| Eggs produced in pullet year. | Number of birds producing in each year the specified number of eggs. | | |
|---|---|---|---|
| | 1899-1900 | 1900-1 | 1901-2 |
| 160 | | | 2 |
| 161 | | 3 | 1 |
| 162 | | | 1 |
| 163 | 1 | 2 | |
| 164 | 1 | | |
| 165 | | 1 | |
| 166 | 1 | 2 | |
| 167 | 1 | 1 | |
| 171 | 2 | | |
| 173 | 1 | 2 | |
| 175 | 1 | | |
| 176 | | 1 | |
| 178 | 2 | | |
| 179 | 2 | | |
| 180 | 1 | | |
| 181 | | 1 | |
| 183 | | 1 | |
| 184 | | | 1 |
| 185 | | | 1 |
| 186 | | 2 | |
| 191 | | | 1 |
| 193 | 1 | | |
| 195 | 1 | 1 | |
| 196 | 1 | | |
| 203 | | 1 | |
| 207 | | 1 | |
| 206 | 1 | 1 | |
| 209 | 1 | | |
| 214 | 1 | | |
| 217 | 1 | 1 | |
| 226 | | 1 | |
| 233 | | 1 | |
| Total.... | 70 | 72 | 33 |

TABLE III.—*Egg production of the same individual hens in the first and second laying years.*

| Eggs laid in— | | Eggs laid in— | | Eggs laid in— | | Eggs laid in— | |
|---|---|---|---|---|---|---|---|
| First year. | Second year. | First year. | Second year. | First year. | Second year. | First year. | Second year. |
| 161 | 61 | 135 | 161 | 195 | 166 | 190 | 88 |
| 175 | 8 | 137 | 168 | 200 | 176 | 170 | 103 |
| 155 | 118 | 165 | 165 | 185 | 147 | 189 | 13 |
| 165 | 112 | 151 | 82 | 209 | 138 | 95 | 82 |
| 175 | 94 | 154 | 94 | 234 | 150 | 177 | 58 |
| 181 | 159 | 120 | 115 | 191 | 73 | 175 | 98 |
| 142 | 89 | 143 | 46 | 99 | 67 | 157 | 86 |
| 191 | 94 | 129 | 125 | 183 | 124 | 167 | 51 |
| 165 | 151 | 191 | 147 | 107 | 22 | 200 | 185 |
| 168 | 40 | 151 | 150 | 61 | 20 | 209 | 102 |
| 140 | 103 | 138 | 148 | 213 | 137 | 170 | 107 |
| 160 | 145 | 208 | 127 | 185 | 106 | 51 | 86 |
| 201 | 30 | 237 | 102 | 92 | 35 | 181 | 70 |
| 136 | 130 | 163 | 71 | 190 | 143 | 47 | 44 |
| 184 | 99 | 211 | 145 | 169 | 30 | 192 | 133 |
| 152 | 162 | 196 | 16 | 74 | 12 | | |
| 155 | 116 | 170 | 172 | 182 | 34 | | |

()